THE GEOGRAPHY OF INNOVATION

Economics of Science, Technology and Innovation

VOLUME 2

The titles published in this series are listed at the end of this volume.

THE GEOGRAPHY OF INNOVATION

by

MARYANN P. FELDMAN
Department of Economics,
Goucher College,
Baltimore, Maryland

KLUWER ACADEMIC PUBLISHERS
DORDRECHT / BOSTON / LONDON

A C.I.P. Catalogue record for this book is available from the Library of Congress.

ISBN 0-7923-2698-9

Published by Kluwer Academic Publishers,
P.O. Box 17, 3300 AA Dordrecht, The Netherlands.

Kluwer Academic Publishers incorporates
the publishing programmes of
D. Reidel, Martinus Nijhoff, Dr W. Junk and MTP Press.

Sold and distributed in the U.S.A. and Canada
by Kluwer Academic Publishers,
101 Philip Drive, Norwell, MA 02061, U.S.A.

In all other countries, sold and distributed
by Kluwer Academic Publishers Group,
P.O. Box 322, 3300 AH Dordrecht, The Netherlands.

Printed on acid-free paper

Printed in the Netherlands

to the memory of my mother

Contents

List of Tables and Figures

Tables

Figures

Preface

This book offers a geographic dimension to the study of innovation and product commercialization. Building on the literature in economics and geography, this book demonstrates that product innovation clusters spatially in regions which provide concentrations of the knowledge needed for the commercialization process. The book develops a conceptual model which links the location of new product innovations to the sources of these knowledge inputs. The geographic concentration of this knowledge forms a technological infrastructure which promotes information transfers, and lowers the risks and the costs of engaging in innovative activity. Empirical estimation confirms that the location of product innovation is related to the underlying technological infrastructure, and that the location of the knowledge inputs are mutually-reinforcing in defining a region's competitive advantage. The book concludes by considering the policy implications of these findings for both private firms and state governments.

This work is intended for academics, policy practitioners and students in the fields of innovation and technological change, geography and regional science, and economic development. This work is part of a larger research effort to understand why the location of innovative activity varies spatially, specifically the externalities and increasing returns which accrue to location.

Acknowledgements

This work has benefitted greatly from discussions with friends and colleagues. I wish to specifically note the contribution of Mark Kamlet, Wes Cohen, Richard Florida, Zoltan Acs and David Audretsch. I would like to thank Gail Cohen Shaivitz for her dedication in editing the final manuscript.

I would like to thank the following journals for permission to include materials that first appeared in their pages. These include the *American Economic Review*, the *Annals of the Association of American Geographers*, the *Economic Development Quarterly*, the *Journal of Corporate and Industrial Change* and *Small Business Economics*. A list of articles can be found in the bibliography.

1
Introduction

While there is general agreement that the rate of technical change is important in determining an economy's rate of growth, we have a limited understanding of the sources of technical progress and of why the pace of progress varies over time and space (Lucas 1993). The new growth theories find that the divergence in growth rates may be a result of increasing returns to knowledge (Romer 1986). Geographic concentrations of knowledge facilitate information searches, increase search intensity and, in general, ease task coordination. The presence of external economies accruing to knowledge creates spatial differences in the distribution of economic activity (Lucas 1988; Grossman and Helpman 1992).

Innovation, perhaps more than any other economic activity, depends on knowledge. New products are the result of a commercialization process which begins with invention, proceeds with product development, and results in market introduction. Commercially-viable product innovations combine scientific and technical knowledge with knowledge of the market. A new product introduction reflects the successful organization and synthesis of these diverse types of knowledge.

For students of economic development from Adam Smith to Karl Marx and Joseph Schumpeter, innovation is seen as the product of entrepreneurs who harness the resources required for innovation, profit and growth. Analysis of innovation has typically been confined to the organizational boundaries of the individual firm, however the notion that the capacity to innovate incorporates external sources of knowledge has gained acceptance. Innovation is perhaps best characterized as an intrinsically uncertain problem-solving process which blends private knowledge with public knowledge (Dosi 1988a). Private knowledge comes primarily from within firms but is also found in industry associations, scientific and professional societies, and networks of related firms and support services (Nelson 1988). Public knowledge is obtained from institutions that support R&D in scientific and technical fields. These are primarily universities but may also include various government science and technology transfer programs. In this view, innovation embodies a broader landscape of social and economic institutions, and relationships than previously conceived.

Geography, in the most fundamental sense, provides organization for the diverse types of knowledge needed for new product commercialization. We expect that knowledge transverses corridors and streets more easily than continents and oceans (Glaeser, Kallal, Scheinkman and Shleifer 1992). The sources of knowledge are embodied in human and institutional form and are less geographically mobile than financial capital (Sweeney 1987). The sources of knowledge, the public and private institutions in a region, form a technological infrastructure. This technological infrastructure promotes knowledge transfer, facilitates problem-solving, and reduces the risks and the cost of innovation. Once in place, the technological infrastructure creates a capacity for innovation. Due to the cumulative and self-reinforcing nature of knowledge, this capacity or core competence become specialized to particular technologies and industrial sectors (Lundvall 1988; Thomas 1985). As a result of these place-specific concentrations of knowledge, technological advance and industrial competitiveness is enhanced.[1] It is in this way that geography plays an essential role in innovation, and in the growth of advanced, capitalist societies.

Innovation in Economic and Geographic Theory

Since Alfred Marshall, the importance of agglomeration, a form of external economy accruing to geographic concentrations of resources, has been noted in the location and organization of industrial activity. Marshall (1949, 152-153) provides an eloquent description of the benefits of location.

> When an industry has chosen a locality for itself, it is likely to stay there long; so great are the advantages which people following the same skilled trade get from near neighborhood to one another. The mysteries of the trade become no mysteries; but are as it were in the air, and children learn many of them unconsciously. Good work is appreciated, inventions and improvements in machinery, in processes and the general organization of the business have their merits promptly discussed; if one man starts a new idea, it is taken up by other and combined with suggestions of their own; and thus it becomes the source of further new ideas. And presently subsidiary trades grow up in the neighborhood, supplying it with implements and materials, organizing its traffic, and in many ways conducing to the economy of its material.

Recent work has built on Marshall's theories by integrating a spatial dimension to economic theory. Krugman (1991a, 1991b) makes a strong case for the regional or spatial specialization of industrial activity based on the advantages of specialized labor pools and related industries, and the presence of knowledge externalities.[2] Arthur (1988, 1990) finds that activity will cluster spatially given the cumulation of historical events or "path-dependence," and locational "lock-in."

The economics literature focuses on the incentives which encourage individual firms to engage in innovative activity. Recently, the idea that innovative output is determined by a set of knowledge inputs has gained acceptance (Nelson and Winter 1982; Freeman 1989; Griliches 1979; Science Policy Research Unit 1972). Dosi (1988b), Lundvall (1988) and Thomas (1985) note that innovation may have a strong geographic dimension due to the specific and cumulative nature of knowledge-based innovative inputs. Jaffe (1989) demonstrates that spill-overs from industrial R&D and university research are geographically mediated, and influence the location of innovative output such as patenting.

Geographers have long been concerned with issues related to the location of innovative activity. Much of the prior work on the location of innovation has implicitly taken the perspective of a company scanning the landscape to find the optimal site. This type of formulation does not recognize the capacity of a region to facilitate or enhance industrial activity. Economic geographers have studied the location of innovative activity (Malecki 1981 1991; Sweeney 1987), the location of high-technology industry (Glasmeier 1988; Hall and Markusen 1985), and the dynamics of regional innovative complexes (Stohr 1986). More recently, others have emphasized the importance of agglomeration economies based upon concentrations of resources and networks (DeBresson and Amesse 1991). There are many case studies of regional innovation complexes, such as Route 128 (Dorfman 1983), Silicon Valley (Saxenian 1985), or Orange County (Scott 1988). These case studies suggest that innovation is a complex geographic process with multiple spatial determinants.

The Innovation Data

A fundamental obstacle to more systematic analysis of the effect of location on innovative activity is the lack of a good measure of innovative output. Griliches (1990: 1669) asserts that "The dream of getting hold of an output indicator of inventive activity is one of the strong motivating forces for economic research in this area." In the absence of direct measures of innovation, alternatives have been used in the literature. These include the amount of inputs used in the innovative process, for example, R&D expenditures (Malecki 1986), or output measures such as the quantity of patented inventions (Jaffe 1989). R&D expenditures indicate resources which are budgeted towards trying to produce innovative output, but do not indicate the success of these efforts. The reliability of patent data is questioned because not all patented inventions prove to be commercially-viable innovations, and many successful innovations are never patented (Mansfield 1984).

This study introduces a direct measure of innovation output which is used to explore the location of innovative activity. In 1982, the Small Business Administration (SBA) conducted a census of innovation citations from over 100 scientific and trade journals. The SBA data capture commercial innovation which, by nature of the citation, add economically useful knowledge to a product category. In contrast to patent data, which mark the certification of a new invention, innovation citations announce the market introduction of a commercially-viable product.[3] Acs and Audretsch (1988, 1990) use this data to analyze industrial dynamics.

There are some limitations to these data, and potential sources of bias to consider. The data are only available as a cross-section for the year 1982. This limits us from investigating any technological and industrial restructuring which may have occurred since that time. The year 1982, however, is a particularly useful year to explore geographic phenomenon related to innovation. The early 1980s are recognized as a period of considerable innovation, with the emergence of new high-technology industries such as personal computers, electronic design automation, software, measuring and controlling, and communication devices. This innovation is accomplished by small high-technology start-up companies as well as larger, established firms, such as IBM and DuPont. In this regard, 1982 provides a particularly useful vantage point from which to examine the geography of innovation.

Another concern is that innovation citations may be biased towards items which the journal editors consider to be of special interest. The editors may have a bias towards unusual products or specific types of products. Also, some product innovations may not be reported because they are simply not of interest to the journal editors. We should further consider that product announcements reflect a variety of organizations with different capabilities and varying motives. This may introduce bias with regard to firm size.[4] For example, large firms with dedicated public relations departments may have greater rapport with journal editors. This may result in an over-representation of innovative activity from large firms. This potential source of bias may be offset by the fact that small firms view new product announcements as inexpensive advertising and, as a result, small firms aggressively pursue new product citations.

Innovations are attributed to the state in which the establishment responsible for the development of the innovation is located. It is anticipated that some innovations are developed by subsidiaries or divisions of companies with headquarters in other states. Since headquarters may announce new product innovations, the SBA data discriminate between the location of the innovating establishment and the location of the innovating entity (Edwards and Gordon 1984). The site responsible for the major development of the innovation is known as the establishment. The parent company or headquarters is known as the entity. For example, Intel Corporation introduced a 16-Bit Micro-Controller (Model Number 8096). The major development was done by a division of Intel in Arizona while Intel's headquarters are in California. In this case, the state of the establishment is Arizona and the state of the entity is California: the innovation would be attributed to the state of Arizona. For our purposes, the state identifier of the establishment is used to investigate the spatial patterns of innovation.

States are not an entirely satisfactory unit of observation for the investigation of spatial phenomenon. In stark contrast to the micro-economics literature in which the unit of analysis is generally accepted to be the firm, there is no such consensus in the geography literature.[5] The analysis of spatial processes is handicapped by the lack of data on what might be considered an ideal unit of observation. Theoretically, spatial processes occur within the boundaries of geographic areas characterized by functional linkages and dependencies (Czmanski and Ablas 1979). However, geographic data are defined and collected within the boundaries of political jurisdictions. Ideally, we would like data at the city or county level, unfortunately, no finer geographic detail exists.

States are the unit of analysis for which data are available. Therefore, this analysis attempts to be sensitive to the fact that states are not the ideal unit of observation.[6] It is our belief, however, that meaningful inference about the geography of innovation can be made at the state level. In addition, states are an important policy-making unit concerned with fostering innovative activity within their borders.

Research Design and Empirical Approach

This study models the effect of concentrations of knowledge on the location of new product innovation using a knowledge-based innovation production function. Adam Jaffe (1989) demonstrates that university and industrial R&D have a statistically significant effect on the number of patents that originate in a state. However, Jaffe does not consider several key inputs suggested by studies of the innovation process (Kline and Rosenberg 1987). Two additional innovative inputs which are considered in this book are the practical knowledge of a technology which is embodied in related industries and the presence of specialized business services which facilitate the process of introducing a new innovation to the market. The process of solving production problems, meeting customer requirements and overcoming various sorts of 'bottlenecks' has become an important factor in the innovation process (Von Hippel 1988). Business services provide knowledge about consumer demand and help position the product. The intensity of use of external producer services correlates highly with realized product innovations (MacPherson 1991). These findings suggest that related industry presence and business services may be two additional innovative inputs relevant to the location of innovative output. Our model considers the importance of geographically-mediated effects of innovative inputs on innovative output.

The empirical work also investigates the location of innovative inputs. According to an emerging perspective, innovation relies on regional innovative capacity which is the result of a cumulative process. This process is based on increasing returns to geographic concentrations of resources, and the external economies of scope generated by co-location of complementary resources.

Summary of Findings

The findings of this study are as follows: First, product innovations exhibit a pronounced tendency to cluster geographically. This confirms a long-held but previously unconfirmed suspicion in the literature (Hall 1985). Second, the geographic-clustering of product innovations at the state level is related to the level of university R&D and industry R&D expenditures in the state. This finding is consistent with Jaffe (1989). Third, this study further suggests that the clustering of product innovation at the state level is related to other innovative inputs, that include the presence of related industry and the presence of specialized business services. This finding is consistent with the view of innovation as a process facilitated by diverse types of expertise and knowledge (Kline and Rosenberg 1987). The presence of these complementary activities promotes information spill-overs which lower the cost of developing new innovations for firms located within these areas. This suggests that firms located in states that contain concentrations of these sources of knowledge which enhance the innovation process, will realize a greater number of innovations. Fourth, the empirical results demonstrate relationships between the R&D inputs. Industry R&D expenditures in a state are positively related to university R&D in that state. Moreover, the expenditures for university research increase in departments which are linked to related industry's manufacturing and R&D activity in the state. In this way, industrial R&D expenditures and university research expenditures are mutually-reinforcing and define an area's expertise. This finding is consistent with geographic self-reinforcing expertise suggested by W. Brian Arthur (1990).

Chapter Outline

This section provides a brief synopsis of the remaining Chapters. Chapter Two develops a conceptual model that considers why location matters to innovative activity. Perspectives are drawn from a variety of academic literatures.

Chapter Three presents the locational patterns of innovation at the state level using the SBA Innovation Citation data. These innovations exhibit a strong tendency to cluster geographically, and this tendency is more pronounced when individual industries are considered. Based on the innovation location quotients, certain states appear to have developed a comparative advantage for innovations in specific industries.

Chapter Four develops a model of innovation based on an innovation production function. Innovative activity in a state is found to be statistically significantly related to the presence of R&D inputs from both university and industry and the presence of related industries and specialized business services.

Chapter Five focuses on regional innovative capacity. The location of innovative inputs are not exogenous to the model of the geography of innovation. Behavioral relationships that govern the location of the innovation inputs are specified. Estimation of a system of equation demonstrates relationships between the R&D inputs. Industry R&D expenditures in a state are positively related to university R&D in that state. Moreover, the expenditures for university research increase in departments which are linked to related industry's manufacturing, and R&D activity in the state.

Chapter Six concludes by considering the policy implications of the findings for business decision-making and for state economic development policy.

Chapter Notes

1. Michael Porter (1989) notes the importance of geographic clustering to furthering international competitive advantage. The ability to produce new product innovation in a timely manner is key to the success of these areas.

2. See David and Rosenbloom (1990).

3. Information as to the economic significance or the revenue generated by each innovation is not available. We know that 80% of the innovations were considered by the editors and product developers as an incremental improvement to an existing product category (Edwards and Gordon 1984). An example of the demands of this type of determination is provided by Trajtenberg (1990) in an extensive study of CT scanners. The SBA data collection effort surveyed editors and company officials to discern the economic significance of each innovation.

4. Although the data was collected by the U.S. Small Business Administration, the purpose was to provide a census of new product innovations introduced to the market in 1982. The reader should not assume that the data are small firm innovations because of the source.

5. Agglomerations have been measured in the literature using a variety of geographic units ranging from the multi-state Manufacturing Belt (Norton and Rees 1979) to the Metropolitan Statistical Area (Oakey 1984).

6. The empirical estimation includes control variables to mitigate the aggregation bias caused by using state rather than a smaller and more uniform geographic unit.

2

Why Location Matters for Innovative Activity

For the past twenty years a prominent view seemed to indicate that geographic location had become less important to economic activity (Barnet and Muller 1974; Vernon 1977). It is well known that as production processes standardize manufacturing industries, even high-tech manufacturing industries, become "footloose" -- seeking out the lowest cost locations (Hymer 1979; Bluestone and Harrison 1982). In contrast, the non-routine production activities associated with innovation such as research and development, experimental and prototype manufacturing, and small volume production are increasingly spatially concentrated (Malecki 1980).

Innovative activity tends to cluster spatially in what is known as the "Silicon Valley" phenomenon. Silicon Valley has become a prominent example of this type of geographic clustering due to the rapid growth of semi-conductors and personal computers, however there is evidence that geographic clustering has been important to technological advance throughout history.[1] This chapter considers why location matters to innovative activity. We focus on how characteristics of the innovation commercialization process interact with place specific assets and expertise to enhance innovation and lead to geographic clustering.

In order to integrate innovation, an industrial process, with geographic location we draw perspectives from several literatures. The industrial organization literature is concerned with the innovation process and identifies important inputs, specifically knowledge inputs. The geography literature focuses on location and the ways in which agglomerations of resources create spill-overs which lower the cost of innovative output. Together, these literatures suggest that higher levels of innovative output will be associated with locations which contain concentrations of knowledge critical to the commercialization process.

Models of the Innovation Process

An examination of the commercialization process provides insights towards understanding the relationship between geography and innovation. For many years, conventional wisdom viewed innovation as a straightforward path from the laboratory directly to the marketplace. This type of abstraction is a limited case and only describes the innovation process under certain, restricted circumstances. The main strength of the linear model is its description of the stages of the process. This first stage is basic research which results in scientific discovery. Next, applied research develops and refines the discovery into a product. The final stage is market introduction which results in the manufacturing and marketing of the product.

Figure 2-1: The Linear Model of Innovation

In the linear model, scientific discovery is the only source of ideas for new products. This, however, contradicts evidence that there are other types of knowledge and expertise that provide ideas for new product innovation. For example, the practical experience which is gained from using a product is an important source of new innovations

(Von Hippel 1988). There are certain aspects of a technology which are tacit and are best understood by working with, and using, that technology. The knowledge gained through a variety of activities outside of the R&D lab create an understanding of the potential of a technology which may provide the ideas for new products. Product users, or customers, have a unique familiarity with a technology and may suggest new products based on the ways in which existing products fail to meet their needs. Some customers may provide specifications which provide ideas for large scale commercial introduction. Other sources of technical information which may precipitate new product innovations may be found from other related firms such as suppliers, distributors and competitors.

Rather than being a final perfunctory end stage, marketing or more broadly, producer services, may provide a catalyst for new product development. Managers' intuitive sense of market trends may provide a spark for new product development. Formal activities, such as marketing research, provide information on customer needs which may suggest product modifications and improvements, and may help to position the product for increased commercial success. Test marketing an innovation can reveal the potential success of a product or can suggest modifications that send a prototype product back to the drawing board. Other specialized producer services provide managerial, accounting, legal, and financial expertise which can reduce the risks and opportunity costs associated with product commercialization. Clearly, these activities provide more critical inputs to the innovation process than the linear model allows.

This simple model does not allow for any flows of information between the stages of the process. Practical examples support the existence of this type of feedback and suggests that synthesis of different types of knowledge characterizes successful innovation. As an example, the Matsushita Bread Machine was developed in co-operation with a chef at the near-by Osaka International Hotel (Nonaka 1991: 98-99). The company R&D department had trouble designing the machine so that it would knead the bread dough evenly and consistently. One of the product developers apprenticed with a chef who had the reputation for making the best bread in the city of Osaka. Through this process, she learned a distinctive way of stretching and twisting the dough. This technique was taken back to the R&D lab and incorporated in the new product design. The result set sales records for a new kitchen appliance.

An often-heard complaint about American product design is that R&D engineers "throw their blueprints over the wall;" thereby, separating themselves from the rest of an organization. Dissatisfaction with this type of approach has increased reliance on alternative product development strategies such as new product development teams. In these arrangements, individuals with different expertise are brought together and the melding of their expertise produces improvement in product design with significant time savings. In this type of arrangement, various expertise and knowledge are easily combined via face-to-face contact.

In general, the linear model oversimplifies the organizational challenges that are inherent in the innovation process. In order to successfully commercialize innovation, firms must assemble a wide variety of expertise and knowledge. These knowledge resources are complementary and create synergies which further innovative efforts (Teece 1980). While the linear model stresses the importance of R&D capacity as the link between scientific discovery and commercial introduction, a more realistic view distributes weight among the different types of knowledge. Each of the various types of expertise are critical to completing the commercialization process; the process may not successful if components are missing.

Kline and Rosenberg (1987) suggest a model which addresses the limitations of the linear model. Their conceptualization adds interdependencies and dynamic learning across the various stages of the innovation process. According to their view, innovation may be initiated at any stage, and tends to be circular rather than sequential.

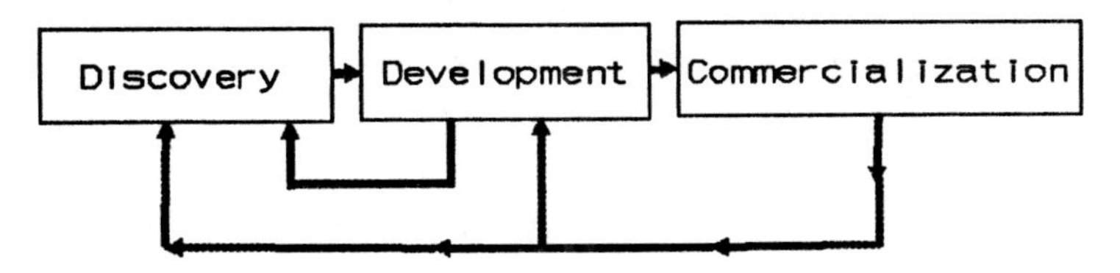

Figure 2-2: Linkage and Feedback Model of Innovation (Kline and Rosenberg, 1987)

Figure 2-2 suggests that the stages need to be linked together and their various expertise melded to secure the success of the innovation. This is the rationale behind the increased use of product development teams. Between firms, we also see examples of cooperative alliances which pool their resources in order to increase innovative potential. While we observe many types of formal strategic arrangements, the same process may occur on a more informal level involving networks of individuals and firms.

Innovative firms must seek out and organize the diverse types of knowledge which facilitate the innovation process. Following a well-known result, firms may be expected to internalize knowledge sources up to the point at which external transactions become advantageous (Coase 1938). External transaction may be preferred if knowledge sources are too costly, too specialized or somehow otherwise constrained from becoming a part of the firm.

To complete the innovation process, firms rely on different sources of knowledge to provide inputs for the stages of the innovation process. Figure 2-3 considers a primary sources of knowledge for each stage of the innovation process model from Figure 2-2. Together these resources constitute a technological infrastructure to support innovative activity.

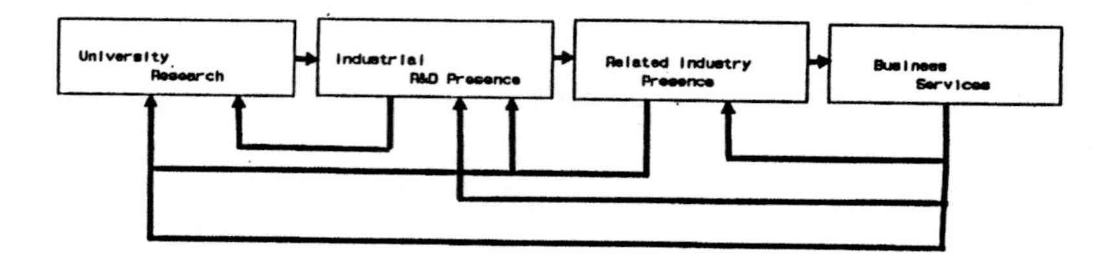

Figure 2-3: Technological Infrastructure of Innovation

The discovery stage relies on scientific knowledge from university research and industrial R&D. Further development and refinement of the innovation is provided by industrial R&D and also by expertise familiar with the technology and gained from manufacturing experience or product use in related industries. Marketing and commercial knowledge is provided by producer services which supply knowledge of the market.

University research is an important source of the basic knowledge which may be important to the innovation process. University R&D enhances the stock of basic knowledge, generates increased technological opportunities across a wide range of industrial fields, and increases the potential productivity of private industrial R&D (Nelson 1986). Mansfield (1991) finds that university R&D generates a social rate of return in excess of 25 percent. Overall, the literature suggests that university R&D tends to have a positive effect on commercial innovation. Jaffe (1989) and Acs, Audretsch and Feldman (1992) identify strong co-location of university and industrial R&D at the state level, and that such co-location of activity has a positive impact on the generation of patents and innovations.

Industrial R&D laboratories provide sources of scientific and technical knowledge required for new product development. Industrial R&D laboratories specialize in market-oriented R&D, specifically, the translation of scientific and technical information into new products. The regional concentration of industrial R&D and its importance to both the innovation process overall and the formation of regional innovation complexes are highlighted in the geography literature (Stohr 1986; Tassey 1991). Malecki (1983) provides extensive evidence of the geographic concentration of R&D by region, state and metropolitan area, with marked concentrations on the East and West coasts. Malecki further (1981) notes regional specialization of R&D, with electronics R&D concentrated in the Boston and San Francisco Bay areas and automotive R&D concentrated around Detroit and Cleveland. Clearly, the consensus is that industrial R&D has a strong positive effect on the location of innovation.

Product development also benefits from an informal knowledge apparatus which is embodied in related industry presence. Some aspects of knowledge have a tacit nature that cannot be completely codified and transferred through blueprints and instructions. This type of knowledge is learned through practice and familiarity in using a technology (Nelson and Winter 1982). Research on high-technology regions suggests that networks of related firms are crucial sources of the new ideas and the sources of the knowledge that contribute to innovation (Stohr 1986; Storper and Walker 1989; Sayer and Walker 1993). Concentrations of firms in related industries provide a pool of technical knowledge, expertise, and other important synergies for the innovation process. For example, product innovations in semiconductors spill-over into related electrical device, consumer electronics, and computer industries. A geographic concentration of related firms provides a potential base of suppliers and users which further develop and refine new innovations. Suppliers and end-users of a technology comprise an important source of outside knowledge and ideas (Von Hippel 1988).

Business services provide market, financial and commercial knowledge. Specialized business services reduce the risks and opportunity costs that are associated with the commercialization by providing knowledge on government regulations and standards, product testing, as well as more traditional services such as financing. These types of specialized producer services tend to locate near their clients (Coffey and Polese 1987). MacPherson (1991) finds that the use of external producer services correlates highly with realized product innovation. The crucial role played by business service providers in high-technology regions such as Route 128 (Dorfman 1983), and Silicon Valley (Saxenian 1985) has been noted.

This model suggests four key knowledge inputs to the commercialization process: networks of firms in related manufacturing industries, concentrations of university R&D, concentrations of industrial R&D, and concentrations of business service providers. The importance of feedback and linkages among the various stages suggests that location may enhance the innovation process by creating greater opportunity for interaction and knowledge dissemination. As knowledge in a product category or industry develops, it becomes cumulative and non-transferable.[2] The knowledge inputs used in commercialization may be embodied in human, institutional, and facility form. These types of resources are relatively immobile and place-specific (Tassey 1991).

Geographers have long recognized that agglomeration economies cause industrial activity to cluster spatially (Weber 1929; Thompson 1968). Agglomeration economies exist when a geographic concentration of resources creates spill-overs which lower the cost of complementary activities. Agglomerations of resources create geographic economies which are external to the benefitting firms. Similar to the more familiar economies of scope realized by complementary activities in large organizations (Teece 1980), agglomeration economies are cost-saving which are realized by geographic proximity. More precisely, the cost of generating two innovations in the same location will be lower than the combined costs of innovating in separate locations:

$$C_1(X,Y) < C_1(X) + C_2(Y). \quad (1)$$

The subscript in the cost functions reference locations 1 and 2.

Agglomeration economies may increase information transfer and promote spill-overs which lower the cost and reduce the risks associated with innovation. The attraction for innovative firms to agglomeration of needed resources is so strong that it is likened to a magnet (Florida and Kenney 1990:54-55). Spill-overs from geographic concentrations of innovative inputs are expected to lower the cost of producing innovative output.[3] The lower cost within these agglomerations suggests that these locations will be advantageous for the production of innovative output. As a result, we expect that these locations will commercialize a greater number of new product innovations.

Innovation as a Cognitive Process

Innovation, at a fundamental level, may be viewed as a communication process that bridges different disciplines with distinct vocabularies and unique motives.[4] While information may be easily transmitted across great distances, translating information into useable knowledge is a more complex, cognitive process. When a technology is relatively stable it can be transmitted in a standardized form such as written articles or correspondence. However, when a technology is new there is great uncertainty about its applications and commercial value. When a technology is more volatile, communication becomes interactive. In the earliest stages, there is typically not even a language to communicate key concepts and there is a need to develop common codes of communication in order to coordinate search procedures. Individuals, especially those with different expertise from diverse backgrounds have different cognitive schemata. Interpreting and synthesizing this information, the process of translating information into knowledge, involves questioning and interpretation. This is a process of trial, feedback, and evaluation that is facilitated by face-to-face interaction.[5] Innovation requires these types of complex reciprocal interactions which result in negotiations and re-conceptualizations as a product moves through the innovation process. There is much information communicated in gestures, facial expression, and tone of voice. Such subtleties provide important clues in the search for meaning and context.

All problem-solving activity uses cognitive models to dictate what information is valuable and how this information should be organized. In order to successfully innovate, it is necessary that scientists be able to communicate with production engineers as well as marketing and financial experts. This requires a shared language and the development of common frames of reference. A set of shared cognitive structures will exist among a group of participants in the innovation process as a common language and similar frames of references are defined. As a result, it becomes possible to assimilate new information into these cognitive structures and for individuals to effectively accumulating the knowledge that facilitates innovation. Only through this cognitive process will interaction promote the learning and knowledge accumulation that contribute to innovative activity.

Machlup (1962) argues that the process of solving a technical problem raises new research questions. Prior experiences determine the ways in which information is interpreted and used. Therefore, opportunities for new innovation build on previous activity in related technologies (Teece 1988). With a developing knowledge of a technology, existing innovators can more easily exploit the commercial value of information and continue to innovate. This activity defines a technological trajectory. Furthermore, there are switching costs associated with moving into new technologies. In this way, knowledge used in the innovation process becomes specific to certain types of transactions and certain types of uses.

Heterogeneity may be critical for innovation. If every person has identical knowledge and background, no amount of communication could increase the stock of knowledge. What you learn from another person depends on what that person knows, not just on how hard you are searching for information. A good example is the Levi-Montalcini and Cohen collaboration which resulted in a Nobel Prize for the discovery of a nerve growth-promoting agent. Rita Levi-Montalcini found that her lack of training in biochemical techniques was an impediment. Then she met Stan Cohen, a biochemist and noted that "The complementarity of our competencies gave us good reason to rejoice instead of causing us inferiority complexes." She recalled that Cohen commented, "Rita, you and I are good, but together we are wonderful (Stephan and Levin 1992: 15)."

David (1992) notes the importance of research communities and complementary expertise in the innovation process. Innovation is more the product of group efforts than the result of solitary genius. An area with innovative activity will develop a set of specialized resources which provide comparative advantage for the next round of innovation. This process is defined by Arthur (1990) as self-reinforcing expertise and gives rise to the geographic clustering of innovative activity.

Stylized Facts on Innovation

Giovanni Dosi provides five "stylized facts" or characteristics of the innovation process which are instructive in further considering why location may benefit innovative activity (1988a). The stylized facts are: the uncertainty of the innovation process; the reliance on university research; the complexity of the innovation process; the importance of learning by doing; and, the cumulative character of innovative activity. Each of these characteristics is considered in turn.

Innovation can be viewed as highly uncertain. This uncertainty extends beyond the lack of information about anticipated events and includes the existence of previously undefined scientific and technical problems. One approach to reduce uncertainty is participation in information exchanges. Innovative networks can be interpreted as the formation of research communities that firms join to exploit new developments in a technology in a timely manner (Nelson 1990). These types of networking activities enable a company to remain at the cutting edge of a technology and to facilitate problem-solving tasks. To the extent that location promotes timely information exchange, innovation and new product commercialization will be enhanced. The cost of membership in innovative networks is a reciprocal sharing of information which creates a de facto market for these transactions. The importance of networking for innovation in specific industries within geographic areas has been documented by Saxanien (1990) for Silicon Valley, and Powell (1989) for the biotechnology industry.[6]

The uncertainty involved in using a new technology provides an incentive for firms to locate together (Lundvall 1988: 355). When technology is standardized and reasonably stable, information exchange may be translated into standard codes. In this case, long distance transmission of information can take place at low costs. On the other hand, when technology is complex and evolving rapidly, long distance standardized transmission is not possible. Therefore, location close to the source of the technology allows firms to translate information into useable knowledge, creating an incentive for firms using complex and dynamic technologies to locate near knowledge sources.

A geographic concentration of rival firms may provide a knowledge resource to reduce innovation's uncertainty. Von Hippel (1988) finds that reciprocal information trading between rival firms provides an important innovative input. A geographic concentration of rival firms appears to facilitate networking and problem-solving, and advances the state of knowledge in the industry (Porter 1990). Allen (1983) suggests evidence of the geographic nature of information trading among rival firms in the nineteenth century English steel industry. As the presence of an industry expands in a given location, firms can specialize in the production of complementary products, and provide expertise to enhance solution searches and further reduce uncertainty.

Universities have become important to the innovation process. Universities emphasize the free exchange and flow of information and their existence in an area creates a sort of intellectual commons. In contrast to the notion that knowledge is a public good easily transferred via publications, gaining commercial control over a new technology requires access to individuals who can turn information into knowledge (Nelson 1989). An example of the importance of face-to-face interaction is provided by a survey of biotechnology researchers by Grefsheim et al. (1991). This work finds that the most important and timely information comes from personal communications which provide information far in advance of printed sources. Researchers reported that the stylistic limitations of formal papers limits their substantive usefulness. Specifically, "Formal papers do not contain the experimenter's strategies and perspectives, nor can they convey what the experimenter thinks the work means and how it dovetails with or contradicts other work" (Grefsheim et al. 1991: 41). While it cannot be disputed that academic conferences and long-distance consulting arrangements provide a means for information dissemination, such contact is less frequent, more costly and qualitatively different.

The complexity of innovative activity increases the scope of the activities needed to complete the commercialization process. To manage this complexity, innovators must conduct intricate search procedures across a variety of disciplines to find specific information. The source of information will be highly specialized within each discipline. The limited usefulness of this information on a day-to-day basis favors external transactions.

The increased scope of innovative activity is suggested by the increased prominence of business services. These services provide information about consumer demand and help shepherd new product innovation through a maze of regulations and product specifications. The specialized services of patent attorneys, market research and feasibility studies, and commercial testing labs will only be internalized by the largest, most sophisticated corporations. Survey work by MacPherson (1991) found that the intensity of the usage of external producer services correlates highly with realized product innovations in medical and chemical firms. Most importantly, since producer services exist solely to supply information, these firms tend to locate near their clients (Coffey and Polese 1987).

There is a certain element of serendipity in the search for relevant information. Shimshoni (1966) argues that the larger the number of skills and interests represented in a given geographical area, the greater the probability of encounters which may lead to fruitful information exchanges. Firms that are located in areas with a range of information sources which could potentially enhance the innovation process, will realize lower search costs in obtaining relevant information.

Some aspects of knowledge have a tacit nature which cannot be completely codified and transferred through blueprints and instructions. This knowledge is learned through practice and practical example (Nelson and Winter 1982). Experimentation in the form of learning-by-doing and learning-by-using play an important role in innovation. Such expertise can come from a variety of sources in related industries. It may be generated by product buyers as they provide information about their needs to enhance product design and development (Von Hippel 1988). This expertise may be facilitated by input suppliers who disseminate technical information which, in turn, facilitates new product innovation (Cohen, Levin and Mowery 1987). In addition, competitors, who face the same obstacles and bottlenecks can be an important source of tacit information (Von Hippel 1988; Porter 1990).

Recent work by Carlson and Jacobsson (1991) suggests that the market for new technologies is primarily regional. The development of technologically-complex products requires close collaboration between suppliers and customers. Until a product becomes standardized, constant specification and design changes make it too costly for suppliers to get involved with distant customers.

In sum, the five characteristics presented here characterize innovation as a process that relies on the timely exchange of information and the accumulation of knowledge. The uncertainty of the innovation process suggests that one way in which firms can reduce uncertainty is by engaging in a reciprocal sharing of information or networking with related firms. The prominence of university research argues for proximity to this innovative input to stay at the cutting edge of technology. The increased complexity of innovation suggests that other sources of information such as related industry presence and specialized business services are key to innovative success. These specialized information sources tend to locate near their client markets. Finally, the cumulative nature of innovative activity suggests that areas with demonstrated innovative success have assembled information that facilitates the next round of innovation. Firms located in areas with limited access to external knowledge inputs must either rely on their own internal efforts or face higher opportunity costs when acquiring external information (Davelaar and Nijkamp 1989; Brody and Florida 1991). Firms may attempt to internalize these knowledge resources by hiring skilled individuals with relevant expertise. But this strategy is geographically bound. The difficult of attracting skilled labor to remote sites is noted by Clark (1981). For the above reasons, innovation is expected to exhibit pronounced geographic clustering. The next chapter turns to a consideration of the empirical examination of the spatial distribution of innovation.

Chapter Notes:

1. For example, see Allen (1983); Landes (1969); Lazonick (1991); Rosenberg (1982).

2. Evidence on the non-transferability of knowledge is provided by Science Policy Research Unit (1972); Thomas (1985); Dosi (1988); Lundvall (1988). Grossman and Helpman (1989) demonstrate this result with a theoretical model.

3. Related empirical work substantiates this conceptualization. Levin and Reiss (1986) conclude that an increase in technological spill-overs reduces unit cost and improves innovative performance. In a study of R&D spill-overs across industries, Bernstein and Naidiri (1988) estimate a .2% average cost saving from a 1% increase in R&D spill-overs.

4. See Carlson and Gorman (1992), Doheny-Farina (1992), and Williams and Gibson (1990).

5. Langlois (1992) documents the development of the computer industry and provides examples of the personal nature of the transfer of technology.

6. Freeman (1991) provides a review of recent studies on the importance of networking in innovative activity.

3

Spatial Patterns of Innovation

Innovation is expected to exhibit strong geographic clustering because new product commercialization relies on knowledge that is cumulative and place-specific. This Chapter explores spatial patterns of innovation using the innovation citation data from the United States Small Business Administration (SBA). The data are based on new product citations and capture innovation which, by nature of the citation, add new, economically-useful knowledge to a product category.

Measuring Innovative Activity

Ideally, we would like to understand, and to measure, the spatial dimension of the economic process that leads to the development of new products. One obstacle to the analysis of the location of innovative activity has been the lack of a direct measure of innovative output. Malecki's studies (1981, 1985, 1986, 1990) examine the location of R&D activities and provide important insights into the factors that influence innovation. R&D expenditures are inputs to the innovation process and represent resources that firms budget toward attempting to produce innovative output, but not necessarily the success of these efforts. Without a measure of innovative output, we cannot determine if R&D expenditures are more productive in certain locations.

The location of patented inventions is examined by Thompson (1962) and Jaffe (1989). Patents provide an indirect measure of innovative output. Griliches (1990), Mansfield (1984), and Scherer

(1983) all warn that the number of patented inventions is not a direct equivalent of a measure of innovative output. Many patented inventions never become commercially-viable products and many successful products are never patented. Scherer (1983) finds the highest propensity to patent, based on the ratio of patents to R&D expenditures, in the following industries: industrial and residential equipment; stone, clay, and glass products; and, household appliances. The high propensity to patent in relatively mundane industries may reflect the incidence of "blocking" patents. Rather than adding economically useful knowledge, these patents may attempt to limit the diffusion of innovation by blocking competition. Mansfield finds that patenting is bypassed in the technology-intensive electronics field because the technical documentation necessary for the patent application releases proprietary design details which may then be exploited by competitors (Mansfield 1984). These problems with patents notwithstanding, the investigation of regional innovative activity has been hindered by the lack of geographic detail on patenting activity.[1]

As a result of the lack of reliable geographic detail on innovative output, employment in high-technology industries is used to study the location of innovation. High-technology industries are defined using measures of R&D intensity. For example, The U.S. Office of Technology Assessment's 1984 study, *Technology, Innovation and Regional Economic Development*, defines innovative industries as those with a ratio of R&D to sales greater than 3.1% and with more than 6.3% of the work-force as scientific and technical workers. Twenty-eight three-digit industries satisfy these criteria, and therefore are defined as innovative industries. The sum of total employment in a state for these industries defines the presence of innovative activity for that state. There are limitations to this measure. The use of employment to define innovation assumes that the employment patterns will be constant across all plants and establishments in an industry. Even though an industry employs a certain percent of scientific and technical workers, these workers are more likely to employed at R&D labs than at branch plants. An employment-based definition gives the greater employment at branch plants more weight than R&D labs. As a result, locations with many branch plants will appear to be relatively innovative even though their activity is dedicated to production.[2]

This research uses a direct measure of innovation output to explore the location of innovative activity. In 1982, the Small Business Administration (SBA) conducted a census of innovation citations from over 100 scientific and trade journals.[3] In contrast to patent data, which marks the certification of a new invention, innovation citations announce the market introduction of a commercially-viable product.[4] The SBA data contains a total of 4,200 manufacturing product innovations with information on the location of the enterprise that introduced the innovation.

Table 3-1: Correlation Matrix for Measures of Innovative Activity[5]

	Innovation	Patents	R&D	Employment
Innovation	1.0000	--	--	--
Patents	.9344**	1.0000	--	--
R&D	.8551**	.8804**	1.0000	--
Employment	.9737**	.9888**	.7013	1.0000

Comparison Among Measures of Innovation

The innovation citation data provide a direct measure of innovative output. How do these data compare to other measures of innovation used in the literature? The innovation citations are highly correlated with other measures of innovative activity. Table 3-1 presents the correlations by state. These high correlations are expected and add

Table 3-2: How Do States Compare on Various Measures of Innovative Activity ?

State	Innovation per 100,000 Workers	Rank	Patents per 100,000 Workers	Rank	% High-Tech Workers	Ran k
Arizona	27.70	8	70.20	19	41	1
California	46.94	3	925.50	1	37	4
Colorado	22.46	9	95.13	16	35	5
Connecticut	28.51	7	312.32	9	39	2
Florida	14.60	18	121.49	14	31	10
Georgia	10.10	27	55.01	22	10	42
Illinois	18.16	13	702.29	4	28	14
Indiana	7.49	34	224.64	10	21	26
Iowa	8.16	32	69.63	20	23	22
Kansas	7.77	33	41.55	24	37	3
Kentucky	3.33	42	59.89	21	17	33
Louisiana	2.39	44	46.70	23	29	13
Massachusetts	51.87	2	383.67	8	33	7
Michigan	11.02	25	494.27	7	15	36
Minnesota	28.65	6	179.94	12	27	16
Missouri	8.20	31	121.20	15	20	27
New Jersey	52.33	1	753.87	3	30	11
New York	29.48	5	818.62	2	22	24
Ohio	15.00	17	572.78	6	22	23
Oklahoma	10.15	26	129.80	13	33	6
Pennsylvania	18.28	12	667.34	5	23	20
Rhode Island	18.46	11	26.93	26	16	35
Utah	11.83	22	29.51	25	31	9
Virginia	9.09	29	85.10	18	26	17
Wisconsin	15.61	16	181.66	11	24	19

validity to the use of the innovation data. These findings indicate that the alternative measures provide reasonable representations of innovation.

Table 3-2 provides a comparison of states for various measures of innovative activity. The alternative measures of innovative activity are highly correlated within states, however, the relative ranking of states between the categories is very different. Employment-based definitions of innovative industries rank states such as Arizona and Connecticut higher than the other measures. This may be attributable to the large number of high-technology branch plants located in these states.[6] The patent measure ranks older, industrial states such as Illinois, Michigan and Ohio higher than the innovation measure.[7] This may be partially explained by examining the industrial mix of these states. Scherer (1983) finds the highest propensity to patent, based on the ratio of patents to R&D expenditures, in industrial and residential equipment, stone, clay and glass products, and household appliances. States with higher concentration of these industries will exhibit relatively more patenting activity. To the degree that innovation citations reflect new commercial innovations introduced to the market, these other measures may misinterpret states' success as sources of innovative activity.

State Patterns of Innovation

While previous research notes that innovative inputs are spatially-concentrated, Peter Hall (1985) finds that the spatial concentration of innovative output has not been demonstrated. We find, using the SBA innovation citation data, that innovative output is highly spatially concentrated. Forty-six states plus the District of Columbia are the location of some innovative output. There is significant concentration of this activity in the eleven states which account for 81% of the 4,200 innovations. The states which produce the greatest number of innovations are California (974), New York (456), New Jersey (426), Massachusetts

Table 3-3: Which States are the Most Innovative?

State	Innovations	Innovations per 100,000 Manufacturing Workers
New Jersey	426	52.33
Massachusetts	360	51.87
California	974	46.94
New Hampshire	33	30.84
New York	456	29.48
Minnesota	110	28.65
Connecticut	132	28.51
Arizona	41	27.70
Colorado	42	22.46
Delaware	15	21.13
National	4200	20.34
Rhode Island	24	18.46
Pennsylvania	245	18.28
Illinois	231	18.16
Texas	169	16.14
Wisconsin	86	15.61
Washington	48	15.38
Ohio	188	15.00
Florida	66	14.60
Oregon	32	14.48

Source: Number of innovations are from the SBA Innovation Citation Data. Number of manufacturing workers are from the *1982 Census of Manufacturers*.

(360), Pennsylvania (245), Illinois (231), Ohio (188), Texas (169), Connecticut (132), Michigan (112) and Minnesota (110). Tables presented in this section will focus on the most innovative states. Complete tables with data for every state are provided in the Appendix.

To normalize for differences in state size, innovation may be measured on a per worker basis. Table 3-3 presents the number of innovations per 100,000 manufacturing employees in 1982; states are ranked from the most innovative to the least innovative. New Jersey has the highest rate of innovation followed by Massachusetts and California. These states generate innovation at greater than twice the national rate of 20.34 innovation per 100,000 manufacturing workers. Seven other states, New Hampshire, New York, Minnesota, Connecticut, Arizona, Colorado, and Delaware were also more innovative than the national average.

Table 3-4 presents the distribution of innovations by two-digit industries and state.[8] The first row presents the total number of innovations for an industry. For example, Food and Kindred Products (SIC 20), had 110 product innovations in 1982. Each cell in the table corresponds to a state and two-digit industry combination. For example, Arizona was the source of twenty-one innovations for Electronic and other Electrical Equipment and Components (SIC 36).

Does geography matter for innovation or is the proportion of innovations attributed to each industry and state simply determined by the industry's and state's propensity to innovate? One hypothesis about the distribution of innovation among states and industries might be that it simply reflects the proportion of innovations attributed to the industry and to the state. For example, the proportion of innovations in electronics in California may be a function of the probability of an innovation occurring in the state of California and the probability of an innovation occurring in the electronics industry.[9] If this were true, location, specifically the interaction of industry with location would have no effect on innovation. In such a case, the distribution of innovations would be deterministic and

Table 3-4: Number of Innovations by State and Industry

State	Selected Two-Digit SIC Code Industries							
	Total	20	28	34	35	36	37	38
National	4200	110	292	201	1392	840	66	1042
Arizona	41	0	0	2	8	21	0	9
California	974	8	18	26	398	288	18	187
Colorado	42	0	0	0	22	9	0	9
Connecticut	132	3	11	5	47	24	3	33
Georgia	53	12	2	3	12	8	1	11
Illinois	231	5	15	18	81	31	1	56
Indiana	49	0	3	2	17	11	2	9
Mass.	360	13	22	15	102	0	3	120
Michigan	112	1	11	7	29	9	21	0
Minnesota	110	6	5	8	38	14	0	28
New Jersey	426	6	75	19	114	58	0	131
New York	456	20	39	20	111	0	2	147
Ohio	188	3	16	21	51	27	3	44
Pennsylvania	245	14	27	16	78	30	0	66
Texas	169	4	14	10	58	35	2	37
Wisconsin	86	1	2	6	28	17	1	29

Note: Complete state data are provided in the Appendix. See Table A-4. SIC 20: Food and Kindred Products; SIC 28: Chemicals and Allied Products; SIC 34: Fabricated Metal Products; SIC 35: Industrial and Commercial Machinery and Computer Equipment; SIC 36: Electronic and Other Electrical Equipment; SIC 37: Transportation Equipment; SIC 38: Measuring, Analyzing and Controlling Instruments geography would not matter for innovation. However, a Chi-Squared test of the independence of state and industry innovations rejects this hypothesis. Therefore, we conclude that the number of innovations in a state and industry is not determined strictly by the state's and industry's propensity to innovate.

Another hypothesis is that the location of innovation is solely a function of the location of industry. Industries are geographically-concentrated and the geographic distribution of innovation may be determined by the geographic distribution of manufacturing. In this case, the number of innovations in an industry and state would be highly correlated with industrial presence in that state. The correlation of the count of innovation and state manufacturing value added was 0.236. For two-digit sectors and states, the correlation between innovation and value-added was .4202. Although innovation is positively correlated with industry presence, the relationship is not deterministic.

The geographic concentration of innovation is more pronounced with greater industry detail. Table 3-5 provides the distribution of innovations by state for a subset of the most innovative three-digit industries. The industries in this table are ranked by the total number of innovations and indicate geographic distribution. For example, there are 954 product innovations in computers (SIC 357) and 668 product innovations in measuring and controlling instruments (SIC 382). Column (3) lists the number of innovations that are attributed to each state. For each three-digit industry, the two states which account for the highest number of innovations are listed. For example, the state of California provided 365 innovations for SIC 357. This represents, on average, one innovation in the computing machinery industry in California for every day of the year. This was 38.3% of the innovations in the computing machinery industry as indicated by Column 4. The state of California has a grand total of 974 innovations. Column 5 reports that the computing machinery industry represents 37.6% of the state's total innovations.

Table 3-5: Distribution of Three-digit Innovations by State

Product	(2) State	(3) Count	(4) % of Innovations	(5) % of State	(6) Location Quotient
Computers SIC 357					n=954
	California	365	38.3	37.6	167.84
	Massachusetts	82	8.6	22.8	100.44
Measuring and Controlling Instruments SIC 382					n=668
	California	134	20.1	13.8	126.42
	Massachusetts	94	14.1	26.1	164.15
Communication Equipment SIC 366					n=376
	California	116	30.9	11.9	132.22
	New York	45	12.0	9.9	110.00
Electrical Components SIC 367					n=261
	California	128	49.0	13.2	211.29
	Massachusetts	26	10.0	7.2	116.13
Medical Instruments SIC 384					n=228
	New Jersey	57	25.0	13.5	248.15
	New York	51	22.4	11.2	207.41
General Industrial Machinery and Equipment SIC 356					n=164
	Pennsylvania	25	15.2	10.2	261.54
	New Jersey	18	11.0	4.2	107.69
Drugs SIC 283					n=133
	New Jersey	52	39.1	12.3	381.25
	New York	18	13.5	4.0	121.88

In the literature, location quotients, in the last column of Table 3-5, assess the degree of geographic specialization or the extent to which an activity is represented in a geographic area compared to its representation in the national economy. The innovation location quotients, found in column 6, are calculated as the percentage of innovation in a state accounted for by an industry divided by the percentage of national innovations accounted for by that industry:

$$LQ_{is} = \frac{\lambda_{is}}{\lambda_{i.}} \times 100. \tag{2}$$

where λ_{is}, the numerator, is the percentage of the innovations in a state, s accounted for by the industry, i. The denominator is the percentage of national innovations accounted for by that industry. The ratio is then multiplied by 100.

An innovation location quotient of 100 indicates that innovation is equally represented in the state and national economies. A location quotient greater than 100 indicates relative specialization in that activity. For the seven most innovative industries listed in Table 3-5, the average location quotient is 218.39, indicating that these areas have achieved some specialized advantage for that industry. Indeed, on average, the most innovative state accounts for 31% of the innovations within an industry. Almost one-third of the innovations in innovative industries are concentrated in the most innovative location.[10]

This descriptive analysis leads one to question why manufacturing product innovations cluster in certain locations. What factors and forces determine the location of innovative output? The next section examines this issue from various perspectives.

Table 3-6: State Expenditures for Industrial R&D and University Research: 1977

State	(1) Industry R&D Funds	(2) University Research Funds	Ratio of (1)/(2)
California	$ 5,600	$ 586	9.6
Connecticut	$ 876	$ 89	9.8
Illinois	$ 1,360	$ 199	6.8
Massachusetts	$ 1,349	$ 298	4.5
Michigan	$ 2,750	$ 171	16.1
Minnesota	$ 602	$ 95	6.3
New Jersey	$ 2,191	$ 68	32.2
New York	$ 2,542	$ 482	5.3
Ohio	$ 1,183	$ 137	8.6
Pennsylvania	$ 1,845	$ 240	7.7

Note: Data for 1977 is used to accommodate the time-lag between R&D expenditures and market introduction. Expenditure funds are in millions of nominal 1977 dollars and are from National Science Board (1989).

The Location of Innovative Inputs

Two important inputs to the innovation process are university R&D and industrial R&D. Table 3-6 presents university research and industrial R&D expenditures for a select group of innovative states. University research and Industrial R&D are presented in millions of nominal 1977 dollars. The last column in Table 3-6 presents the ratio of industry R&D expenditures to university research expenditures.

This presentation is instructive for further analysis. On the basis of expenditures alone, large population states like California predominate. This suggests that further analysis should accommodate for population size in order to facilitate cross state comparisons. The year 1977 is chosen to approximate the time-lag between product development and commercial introduction. The use of funds for a single year raises questions about yearly fluctuations in expenditure levels. The innovative capacity of an area may reflect the accumulated stock of the innovative inputs rather than a single one-year flow. This suggests that the analysis use the stock of input expenditures over a longer time period. In addition, the more pronounced clustering within industry groups suggests that the analysis should be conducted at the industry level.

Table 3-7: Number of Innovations Yielded Per Unit of Innovative Input for the Computer Industry, SIC 357: Selected States

State	(1) N	(2) University R&D		(3) Industrial R&D		(4) Related Industry	
		Funds	Ratio	Funds	Ratio	Value-Added	Ratio
California	365	$366.0	1.00	$3883	0.094	$4195.84	0.086
Connecticut	29	$ 7.2	4.03	$ 650	0.045	$1037.72	0.028
Florida	17	$ 10.6	1.60	$ 375	0.045	$ 515.03	0.033
Illinois	35	$124.2	0.28	$ 894	0.039	$4404.49	0.008
Massachusetts	82	$119.6	0.69	$ 954	0.085	$2117.67	0.039
Michigan	14	$ 21.8	0.64	$1815	0.007	$3050.29	0.005
Minnesota	24	$ 6.4	3.75	$ 399	0.060	$1444.81	0.017
New Jersey	65	$ 40.5	1.60	$ 1361	0.047	$1242.57	0.052
New York	77	$ 88.5	0.87	$1859	0.041	$3161.09	0.024
Ohio	16	$ 20.3	0.79	$ 926	0.017	$3842.52	0.004
Pennsylvania	34	$ 34.3	0.99	$1293	0.026	$2518.65	0.013

Ratio refers to the number of innovations per unit of innovative input. University Research and industry R&D funds and industry value-added are in millions of 1972 dollars.

Table 3-7 examines the relationship between innovative output and innovative inputs for the computer industry (SIC 357). For this descriptive analysis the innovative inputs are measured as the average annual expenditure over the ten-year time period prior to the introduction of the innovations in 1982. Expenditures are in millions of 1972 real dollars. Column (1) shows the number of innovations by state. Column (2) presents both university R&D expenditure and the ratio of the number of innovations yielded per dollar of university research expenditure. For example, one million dollars of university R&D in California produced one product innovation in the computer industry. Column (3) represents the total state industrial R&D funds along with the ratio of the number of innovations per dollar of industrial R&D expenditure. University expenditure yields more innovations per dollar spent because of lower expenditure levels. By way of example, the California data indicates that $11 million dollars in industrial R&D yields one product innovation in the computer industry.

Additionally, Table 3-7 includes a measure of related industry presence in the state. Related industry is measured as value-added for the 2-digit industry 35. To capture the stock of related industry technical information and avoid yearly fluctuations, the ten-year average in millions of real dollars is used. Value-added for SIC 35 is positively correlated with product innovations for SIC 357. This description examines the location of three innovative inputs: university research expenditures; industrial R&D expenditure; and related industry presence, and finds a correspondence between the concentration of the innovative inputs and the presence of innovation output at the state level. A test of the relationship between the concentration of innovative inputs and the location of innovative output requires further empirical examination.

Real Effects of Academic Research

Adam Jaffe (1989) provides the first examination of the extent to which spatially-mediated R&D spill-overs influence the generation of increased innovative output at the state level. While prior research considers R&D spill-overs across technical areas, this work reflects the first attempt to model geographically-mediated spill-overs.[11] The results demonstrate that patenting activity increases in the presence of high corporate R&D expenditures. In addition, corporate patent activity responds positively to knowledge spill-overs from university research. This evidence suggests that industrial R&D and university research exhibit geographically-mediated influences on the number of commercial patents. The amount of R&D expenditures in a state has a positive and significant effect on the number of corporate patents which originate in that state.

Jaffe uses patents to identify the contribution of university research in generating "new economically-useful knowledge." As discussed earlier, patents are not a direct measure of innovative output. In contrast, the innovation data exclude those inventions which were patented but did not prove to be viable for market introduction and include innovations which were never patented. Are the same effects realized if the innovation citation measure is substituted into the Jaffe's model?

Jaffe's model uses a knowledge production function framework with a two-input modified Cobb-Douglas specification:

$$\begin{aligned} logP_{is} = {} & \beta_1 logIND_s + \beta_2 logUNIV_{is} \\ & + \beta_3 logPOP_s + \beta_4 (logUNIV_{is} * logC_s) + \epsilon_{is} \end{aligned} \tag{3}$$

The unit of observation is at the level of the state, *s*, and the "technological area", or industrial sector, *i*. Jaffe assigns patents to the five technical areas of drugs, chemicals, electronics, mechanical arts, and all others. For innovative output, Jaffe uses the number of corporate patents per technical area in a state. The two innovative inputs are total industrial R&D expenditures by state, $IND_{.s}$, and university research expenditures by technical area by state, $UNIV_{is}$. Unfortunately, industrial R&D expenditures are not available at the industry level. In order to compensate for using state level data to measure geographic spill-overs, Jaffe constructs an index of the geographic coincidence of university and industrial research labs, $C_{.s}$.[12] State population, $POP_{.s}$, is included to control for size differences across the geographic units of observation.

Table 3-8 : Summary Statistics for Regression Sample using Data from Jaffe (1989) and Innovation Citations

	Mean	Standard Deviation	Minimum	Maximum
University Research	98.8	144.0	12.0	710.4
Industrial R&D	582.9	823.5	3.8	4328.9
Geographic Index	0.63	0.35	0.03	1.00
Population	5955.9	4853.0	946	24,265
Patents	879.4	975.7	39.0	3230.0
Innovations	130.1	206.4	4.0	974.0

University research and Industrial R&D expenditures are reported in millions of 1972 dollars. Patents and innovation are the count for each measure. Population is reported in thousands.

Table 3-8 compares the mean measures of university research expenditures and corporate patents used by Jaffe against the mean number of innovations per state. University research and industrial R&D expenditures are presented in millions of 1972 dollars.[13] Jaffe pools eight years of data across states in estimating the production function for patents. This is not possible with the cross-sectional innovation data. Further, it is important to recognize that Jaffe's results do not vary greatly when the model is estimated using the eight-year mean values. Table 3-9 provides a comparison of Jaffe's results using the eight-year pooled patent data against results using the eight-year mean.

Not surprisingly, the mean patent measure yields virtually identical results to the results based on the pooled estimation. That is, both private corporate expenditures on R&D, and expenditures by universities on research are found to exert a positive and significant influence on patent activity. The coefficient on industrial R&D is remarkably close to the coefficient estimated using the pooled sample. It should be noted that the coefficient on university R&D is larger and more statistically significant in the mean patent model. However, the use of mean values do not change the conclusion.

Table 3-9: Comparison of Regression Results Using Jaffe's Pooled Patent Measure and the Mean Patent Measure

	Pooled Patent Measure	Mean Patent Measure
$Log(IND_{s})$	0.713[a] (0.035)	0.668[a] (0.078)
$Log(UNIV_{is})$	0.084[b] (0.047)	0.241[a] (0.066)
$Log(UNIV_{is})*log(C_{s})$	0.109[a] (0.041)	0.020 (0.082)
$log(POP_{s})$	0.179[a] (0.068)	0.059 (0.046)
n	1160	145
R^2	0.915	0.959

Standard Errors in Parenthesis. Note: [a] indicates significance of at least .95; [b] indicates significance of at least .90.

Table 3.10 provides a comparison of the results using both the patent measure and the innovation measure. Substitution of the direct measure of innovative activity for the patent measure in the knowledge production function generally strengthens Jaffe's arguments, and reinforces his findings. The effect of university spill-overs is apparently greater on innovations than on patented inventions. The elasticity of the $log(UNIV_{is})$ almost doubles from 0.241 when the patent measure is used in comparison to 0.431 when the innovation measure is used. Most importantly, the innovation data provides even greater support that spill-overs are facilitated by geographically-mediated spill-overs from university research and industries R&D activities. Clearly, with both of these measures of innovative output, we can conclude that industry R&D and university research have a strong and statistically significant geographically-mediated effect.

Table 3-10: Comparison of Regression Results Using Jaffe's Patent Measure and the Innovation Measure

	Patents	Innovations
Log($IND_{.s}$)	0.668^{a} (0.078)	0.428^{a} (0.093)
Log($UNIV_{is}$)	0.241^{a} (0.066)	0.431^{a} (0.071)
Log($UNIV_{is}$)*log($C_{.s}$)	0.020 (0.082)	0.173^{b} (0.090)
log($POP_{.s}$)	0.059 (0.046)	-0.072 (0.056)
n	145	125[14]
R^2	0.959	0.902

Standard Errors in Parenthesis. Note: [a] indicates significance of at least .95; [b] indicates significance of at least .90.

Differences between the two measures may reflect two separate and distinct stages in the innovation process. Patents represent new inventions, not new commercial introductions. As we might expect, industry R&D expenditures have a larger coefficient and statistically stronger effect on patents than on innovations. Because patents are closer to the initial stages of the innovation process, they may be more directly related to the work of the R&D lab. On the other hand, university research has a greater effect on the number of commercially-viable innovations in a state. This result is somewhat surprising because university R&D is generally thought to be typically far removed from the market. The specification might consider additional influences which account for the location of innovation.

Commercial innovation may benefit from spill-overs of other types of activities that are omitted from this two-input model. Commercial innovation is the result of several different, but complementary types of knowledge; that is, scientific and technical knowledge coupled with practical knowledge of the market. Technical knowledge, partly codified and formal, and partly informal and tacit, provides the discovery and development of a new invention. In order to generate a profit, this knowledge must be coupled with a complementary knowledge of the market and an understanding of consumer demand. University research and industrial R&D are important inputs to the invention process which result in the granting of a patent, however other types of knowledge inputs are useful in the commercialization process.

The commercialization process is more complicated than the patenting process, and spill-overs from other inputs may effect the location of product innovation. Most importantly, innovation may benefit from spill-overs from other firms that use similar technologies as well as from spill-overs from producer services which provide knowledge relating to consumers and the market. These issues are examined in the next chapter.

Chapter Notes:

1. The computerization of the U.S. Patent office provided Jaffe (1989) an opportunity to look at regional patenting activity. Also see Jaffe, Trajtenberg and Henderson (1993).

2. Branch plant locations are more likely to engage in process innovation (Howells 1990).

3. This data was analyzed in considerable detail by Acs and Audretsch (1988,1990). The data are introduced on page 4. A more complete description of the data is provided in Appendix A.

4. Information as to the economic significance or the revenue generated by each innovations is not available. An example of the demands of this type of determination is provided by Trajtenberg (1990) in an extensive study of CT scanners.

5. Patent counts by state are from Jaffe (1989) and represent the average annual number of patents received in 29 states over an eight-year period. High-technology employment data are from the U.S. Office of Technology Assessment (1984) for the year 1982 for the ten states with the highest high-technology employment levels. R&D expenditures are from the National Science Foundation as reported by Jaffe (1989). An alternative employment-based measure of innovations is high-tech employment as a percentage of non-agricultural employment (Glasmeier, 1985). This measure had a correlation coefficient of .4474 with innovations and .3201 with patents.

6. Estimates of employment in high-technology industry used in this table are from Glasmeier (1985). This measure is used because the counts of high-technology employment used in table 3-1 are available for only ten states. This measure has a correlation coefficient of .4474 with innovations and .3201 with patents.

7. Patent counts are from Jaffe (1989) and represent the average annual corporate patenting activity for a state for the years 1972-1977, 1979 and 1981. Jaffe only provides data for 29 states, and the ranking in Table 3-2 reflects the relative position out of the 29 cases.

8. This is a very abbreviated summary of the complete table found on page 122 of the Appendix.

9. If s references the state and i references the industry, then $p_{is} = p_{i.}p_{.s}$.

10. These results hold for a larger set of industries. See Table A-5 in the Appendix.

11. Zvi Griliches (1992) provides a review of empirical literature on R&D spill-overs.

12. The geographic coincidence index is calculated as

$$C_{s} = \frac{\sum_{c} UNIV_{ic} TP_{ic}}{[\sum_{c} UNIV_{ic}^{2}]^{1/2} [\sum_{c} TP_{ic}^{2}]^{1/2}};$$

where TP_{ic} is the total number of R&D lab workers in a city or SMSA. The geographic coincidence index is calculated as the uncentered correlation of the vectors U_i and TP_i across SMSA's within a state. Jaffe's hypothesis is that research will yield more innovative activity if university and industrial labs are geographically-concentrated. For example, more patents would be expected in Illinois where industrial and research labs are concentrated in Chicago than in Indiana where university labs are located in different SMSA's than industrial labs. For estimation purposes, the $\log(C_{.s})$ is re-scaled to have a zero mean.

13. All data sources and a detailed description of the data and measures can be found in Jaffe (1989). Jaffe uses an eight-year sample (1972-1977, 1979 and 1981) and the data presented in Table 3-8 represent the eight-year average.

14. The number of observations for patents, P_{is}, is the product of 5 technical areas, *i*, and 29 states, *s*. For comparability with Jaffe (1989), the number of innovation observations excludes twenty zero cases. The zero cases are included in Chapter 4 when these results are extended using a Tobit analysis.

4
Technological Infrastructure

New products are the result of a commercialization process which begins with invention, proceeds with development, and concludes with market introduction. This process requires diverse types of knowledge and expertise. In order for an innovation to be commercialized, university and industrial R&D, the inputs identified in the previous chapter, must be combined with a body of practical, technical knowledge as well as a working knowledge concerning consumer demand and the marketplace. This conceptualization suggests that innovation will be advanced by four types of institutions and resources: university R&D, industrial R&D, agglomerations and clusters of firms in related industries, and networks of business-service firms. Taken together, these complementary institutions provide resources and knowledge inputs to the innovation process, generate positive externalities and spill-overs which lower the cost of developing new innovations, and reduce the risks associated with innovation. Once combined, these resources form a technological infrastructure conceptualized as an integrated and spatially-concentrated network of institutions that provide inputs to the innovation process.

This chapter offers a model of a regional technological infrastructure that is based on the diverse types of knowledge which are required to complete the commercialization process. In operational terms, the technological infrastructure is defined in terms of four classes or factors: agglomerations and networks of firms in related industries;

concentrations of university R&D; concentrations of industrial R&D; and networks of business-service firms. The empirical results confirm our conceptualization. Innovation is found to cluster geographically in areas that contain concentrations of specialized resources which comprise the technological infrastructure.

Innovation Inputs

The innovation process model, discussed in the previous chapter, suggests four key knowledge inputs to the commercialization process. The discovery stage relies on information from university research and industrial R&D. Further development and refinement of the product innovation is provided by expertise familiar with the technology and gained from manufacturing experience or product use. Knowledge of the market and the process of bringing an innovation to the market is provided by producer services.

University research is an important source of basic knowledge that may be important to the innovation process. Nelson (1986) and Mansfield (1991) note that university R&D enhances the stock of basic knowledge, generates increased technological opportunities across a wide range of industrial fields, and increases the potential productivity of private industrial R&D. Mansfield (1991) finds that university R&D generates a social rate of return in excess of 25 percent. Overall, the research literature suggests that university R&D tends to have a positive effect on commercial innovation. Malecki (1981) identifies the concentration of university R&D in a few major clusters such as the Boston-Cambridge area and the San Francisco Bay area. Others (Jaffe 1989; Acs, Audretsch and Feldman 1992) identify strong co-location of university and industrial R&D at the state level, and that such co-location of activity has a positive impact on the generation of patents and innovations.

Industrial R&D laboratories provide sources of scientific and technical knowledge required for new product development. With few notable exceptions, such as ATT and IBM which conduct advanced basic research similar to that conducted in universities and government laboratories, industrial R&D laboratories specialize in market-oriented R&D, more specifically, the translation of scientific and technical information into marketable products. The regional concentration of industrial R&D and its importance to both the innovation process overall and the formation of regional innovation complexes is highlighted in the geography literature (Stohr 1986; Tassey 1991). Malecki (1983) provides extensive evidence of the geographic concentration of R&D by region, state and metropolitan area, with marked concentrations on the East and West coasts. Further, Malecki (1981) observes a regional specialization of R&D, with electronics R&D concentrated in the Boston and San Francisco Bay areas and automotive R&D concentrated around Detroit and Cleveland. Clearly, the consensus is that industrial R&D has a strong, positive effect on the location of innovation.

Product development also benefits from an informal knowledge apparatus which is embodied in related industry presence. Some aspects of knowledge have a tacit nature that cannot be completely codified and transferred through blueprints and instructions. This type of knowledge is learned through practice by using a technology, and is best transferred through practical demonstration and usage (Nelson and Winter 1982). Research on high-technology regions suggests that networks of firms, particularly manufacturing firms, are crucial sources of the new ideas and sources of the knowledge that contribute to innovation (Stohr 1986; Storper and Walker 1989; Sayer and Walker 1993). Concentrations or agglomerations of firms in related industries provide a pool of technical knowledge and expertise, and provide a potential base of suppliers and users that further develop and refine these new innovations. Additionally, suppliers and end-users of a technology comprise a important source of outside knowledge and ideas (Von Hippel 1988). Concentrations of firms in related industries create important synergies for the innovation process. For example, product innovations in semiconductors spill-over into related electrical devices, consumer electronics and computer industry.

Business services provide marketing and commercialization information and experience. Specialized business services reduce the risks and the opportunity costs that are associated with the innovation process by providing knowledge on accounting, government regulations and standards, marketing, product-testing, and financing. These types of specialized producer services tend to locate near their clients (Coffey and Polese 1987). MacPherson (1991) finds that the use of external producer services correlates highly with realized product innovation. The pivotal role of business services has been established in the development of high-technology regions such as Route 128 (Dorfman 1983), and Silicon Valley (Saxenian 1985).

In conclusion, the model developed here suggests four key innovative inputs: networks of firms in related manufacturing industries; concentrations of university R&D; concentrations of industrial R&D; and concentrations of business service providers. The presence of these four inputs constitute an area's technological infrastructure. To test the relationship between the technological infrastructure and the location of innovative output requires further empirical examination.

Empirical Model

Following Griliches (1979), innovation is analyzed in a modified production-function approach; innovative output depends on the presence of innovative inputs. The inputs are suggested in the previous chapter by disaggregating the innovation process into the interdependent stages of basic research, applied research, and commercial manufacturing and marketing (Kline and Rosenberg 1987). This perspective suggests that institutions and institutional capabilities may provide outside resources as innovative inputs.

Innovative output, INN_{is}, the number of innovations for an industry, i, and a geographic area, s, is modeled as a function of four innovative inputs:

$$INN_{is} = UNIV_{Is}^{B_1} \, IND_{Is}^{B_2} \, RELPRES_{Is}^{B_3} \, BSERV_{Is}^{B_4}. \tag{4}$$

IND_{Is} is industrial R&D expenditures in the geographic area, s, for the related, larger industry group, I. The more encompassing industry group, *I*, measures the technological area across which spill-overs would be expected to occur. $UNIV_{Is}$ is university research expenditures measured at the level of the academic department and allocated to the relevant industry group. $RELPRES_{Is}$ is the presence of related industry, including the presence of firms in industries using related technologies as well as downstream users of a technology who may disseminate information about the technology. $BSERV_{Is}$ is the presence of specialized business services related to innovative activity within the industry.

Data and Estimation Issues

There are a number of issues to consider with the model estimation. To capture the stock of knowledge inputs, and to avoid yearly fluctuations, the inputs are measured as the ten-year average of annual expenditures for the ten years prior to the market introduction. The dependent variable in equation 4 is the number of innovations for an industry, i, which originated in state, s. The number of innovations by state and industry is a censored-dependent variable; the number of innovation will either be zero or some positive integer. Cases in which there is no innovation output provide information about how innovative locations differ from non-innovative locations. Therefore, including these cases improves the accuracy of the estimation. This analysis uses the Tobit model due to the censored nature of the dependent variable.

Our unit of observation is the state and industry. There were a total of 95 three-digit industries which contain at least one innovation citation. A large number of zero cells result when the data are stratified by state. In order to proceed with the estimation, it is necessary to limit the sample.

Table 4-1: Innovative Three-Digit Industries: Ranked by Number of Innovations

SIC	Industry	Number of Innovations
357	Computer and Office Machinery	954
382	Measuring and Controlling Instruments	668
366	Communications Equipment	376
367	Electronic Components	261
384	Surgical, Medical and Dental Instruments and Supplies	228
356	General Industrial Machinery and Equipment	164
283	Drugs	133
355	Special Industrial Machinery	116
349	Miscellaneous Fabricated Metal Products	105
362	Electronic Industrial Machinery	74
386	Photographic Equipment and Supplies	61
282	Plastic Materials and Synthetic Resins	51
284	Soap, Detergents and Cleaning Preparations	50

The analysis is based on the thirteen most innovative three-digit industries listed in Table 4-1. These thirteen industries account for eighty percent of the total innovations. Each industry included in the estimation contains a minimum of fifty innovation citations. As such, this group is appropriate to study the location of innovative industries.

The SBA innovation data provide information on the state of the establishment that is responsible for developing the innovation. Because the phenomenon we model may be more local in nature, two variables are added to control for aggregation bias. State population, $POP_{.s}$, is included to control for differences in state size and to facilitate cross-state comparisons. Geographic concentration, $CONC_{.s}$, is added to control for within-state variation.[1] One additional control variable, industry sales, $SALES_{i.}$, is included as a control for industry demand which may affect the number of innovations within an industry.

The resulting equation for estimation is

$$\begin{aligned} Log(INN_{is}) = {} & \beta_1 \log(UNIV_{Is}) + \beta_2 \log(IND_{Is}) + \beta_3 Log(RELPRES_{Is}) \\ & + \beta_4 \log(BSERV_{s}) + \beta_5 CONC_{s} + \beta_6 \log(POP_{s}) \\ & + \beta_7 \log(SALES_{i.}) + e_{is}. \end{aligned} \tag{5}$$

Each of the independent variables will be discussed in turn.[2]

$UNIV_{Is}$ represents university R&D expenditures. University research is an important source of the basic knowledge that is expected to be important to the innovation process. Since the dependent variable measures innovation brought to the market in 1982, university research which contributed to these innovations was conducted years earlier.[3] The precise year is difficult to discern because there is great variation in the path to market. Indeed, a realized innovation is most likely the result of the cumulative effect of research that has been conducted over time. To capture the effect of university research expenditures on innovative activity, the average annual expenditures in the ten-year period prior to the introduction of the innovation was taken. Unfortunately, data are not available for the years 1978 and 1980. Consequently, the variable is defined as the eight-year average over the ten-year time period.

Table 4-2: Linking Industries to University Departments	
Industry	University Department
SIC 28: Chemicals and Allied Products	Medicine, Biology, Chemistry and Chemical Engineering
SIC 34: Fabricated Metal Products	Mechanical Engineering and Other Engineering and Physical Sciences
SIC 35: Industrial and Commercial Machinery and Computer Equipment	Electrical Engineering, Astronomy, Physics, Computer Science, Mechanical Engineering and Other Engineering and Physical Sciences
SIC 36: Electronic and Other Electrical Equipment and Components	Electrical Engineering, Astronomy, Physics, and Computer Science
SIC 38: Measuring, Analyzing and Controlling Instruments	Medicine, Biology, Electrical Engineering, Astronomy, Physics, Computer Science, Mechanical Engineering and Other Engineering and Physical Sciences.

There are great differences in the scope and commercial applicability of university research undertaken in different fields. Academic research will not necessarily result in useful knowledge for every industry, however scientific knowledge from certain academic departments is expected to be important for specific industries. To capture the relevant pool of knowledge, academic departments were assigned to relevant industrial fields at the level of the two-digit industry. For example, product innovation in drugs (SIC 283) is linked to research in medicine, biology, chemistry and chemical engineering. Table 4-2 presents the assignment of research funds by academic department to innovative output in an industry. The data is from the National Science Foundation's (NSF) Survey of Science Resources.[4]

IND_{ls} measures the industrial R&D expenditures performed within companies as reported by the National Science Foundation's Science Resources Survey. These data do not include R&D expenditures which are contracted to outside organizations such as universities and colleges,

Table 4-3: Industrial R&D Data Availability

With R&D Data	Innovations	Without Data	Innovations
Alabama	5	Alaska	0
Arizona	41	Delaware	15
Arkansas	5	Hawaii	0
California	974	Idaho	6
Colorado	42	Maine	4
Connecticut	132	Maryland	28
Florida	66	Missouri	36
Georgia	53	Montana	0
Illinois	231	Nevada	1
Indiana	49	New Hampshire	33
Iowa	20	New Mexico	3
Kansas	15	North Dakota	0
Kentucky	9	Oregon	32
Louisiana	5	South Carolina	18
Massachusetts	360	South Dakota	0
Michigan	112	Tennessee	20
Minnesota	110	Texas	169
Mississippi	4	Vermont	6
Nebraska	9	Washington	48
New Jersey	426	West Virginia	4
New York	456	Wyoming	0
North Carolina	38		
Ohio	188		
Oklahoma	20		
Pennsylvania	245		
Rhode Island	24		
Utah	11		
Virginia	38		
Wisconsin	86		

nonprofit organizations, research institutions and other companies. The data represent the ten-year average for R&D expenditures in real dollars.

The sample is defined in the following way. The confidentially requirements of the NSF data suppress data for some locations. As a result, industrial R&D expenditures are available for twenty nine states. The available states, listed in Table 4-3, contain over 92% of the total innovations introduced in the U.S. in 1982. This sample accounts for 78 percent of the U.S. population and 81 percent of the university research expenditures in 1982. Conversely, states for which R&D data is unavailable account for 325 innovations or 7.7 percent of the total innovations.[5] The model estimation uses data for these 29 states.

The question arises as to what is the technological neighborhood across which R&D spill-overs occur. Do R&D spill-overs occur within a narrow band of related industries or do R&D spill-overs cross the entire spectrum of manufacturing industries? This is an empirical question that suggests two alternative specifications of industrial R&D. The first specification uses R&D for the set of related industries, IND_{ls}:

$$\begin{aligned}\log(INN_{is}) = {} & \beta_1 \log(UNIV_{ls}) + \beta_2 \log(IND_{ls}) + \beta_3 Log(RELPRES_{ls}) \\ & + \beta_4 \log(BSERV_{.s}) + \beta_5 CONC_{.s} + \beta_6 \log(POP_{.s}) \\ & + \beta_7 \log(SALES_{i.}) + e_{is}. \end{aligned} \tag{6}$$

The second specification measures total industrial R&D for the state, $IND_{.s}$:

$$\begin{aligned}\log(INN_{.s}) = {} & \beta_1 \log(UNIV_{ls}) + \beta_2 \log(IND_{.}s) + \beta_3 Log(RELPRES_{ls}) \\ & + \beta_4 \log(BSERV_{.s}) + \beta_5 CONC_{.s} + \beta_6 \log(POP_{.s}) \\ & + \beta_7 \log(SALES_{i.}) + e_{is}. \end{aligned} \tag{7}$$

The second specification argues for measurement that is compatible with the data which is available for university research expenditures. Unfortunately, before 1987, the National Science Foundation only provides data on total industrial R&D expenditures at the state level. To test the importance of R&D expenditures for a specific industry, it is necessary to construct estimates of state R&D expenditures at the industry level. Detail on R&D expenditures at the two-digit industry are available from the 1987 National Science Foundation's Survey of Industrial Research and Development. Expenditures are classified under the SIC code of the performing company. The 1987 data was used to allocate the average state R&D expenditures to two-digit industries within a state, IND_{Is}.

The allocation method described above assumes that the percentage of industrial R&D originating within an industry and state has remained relatively stable. There is no evidence that the within-state allocation of R&D expenditures has changed. Malecki and Bradbury (1991) report that the geographic location of corporate R&D facilities in the U.S. has been remarkably consistent since 1970. Glasmeier (1988) notes a movement of scientists and engineers to the South and West during this time period, attributing this migration to the establishment of technological branch plants which locate R&D activity nearer the locus of production. Howells (1990:138) suggests that these units are more oriented towards process innovation and incremental improvements and are fundamentally different from the central R&D labs which are more likely to be involved with product innovation. Nationally, the distribution of R&D expenditures to the industries considered here were stable from 1978 to 1987. The within-state R&D expenditures allocations are assumed to be similarly stable.

$RELPRES_{Is}$ is the concentration of firms in related industries. Concentrations or agglomerations of related firms provide a pool of technical knowledge and expertise, and a potential base of suppliers and users which further refine and contribute to new innovations. Related industry presence is measured as value added for the major industry group related to the industry under consideration. The SIC code classification scheme is designed to accommodate this type of relationship.[6] A three-digit industry is related to other industries within its two-digit major industrial group. For example, drugs (SIC 283), would benefit from the presence of related activity in the industrial group Chemicals and Allied Products (SIC 28). To capture the stock of manufacturing knowledge in a related industry, we use the average real value-added for the corresponding two-digit industry over the ten-year period prior to the introduction of the innovation.

$BERV_{.s}$ represents business services, the last input in the model. The presence of business services related to innovative activity is the most challenging to measure. There are a variety of business services that provide essential knowledge of the marketplace and other aspects of the commercialization process. A case in point is that the services of patent attorneys may be a critical input to the innovative process. Unfortunately, data of this detail do not exist. All legal services are grouped together in SIC 8111 without any finer detail. Additionally, of all the producer-services available to support innovative activity, the only category that is specifically targeted to the introduction of new innovations is commercial testing laboratories, SIC 7397. This is used as a rough measure for business services related to innovative activity. The presence of business services related to innovative activity will be measured as the annual average receipts of Commercial Testing Laboratories over the time period 1972 to 1982, in constant dollars. Table 4-4 presents summary statistics for the data used in the model estimation.

There are three statistical issues that require attention in the model estimation. These include the censored nature of the dependent variable, the likely structure of the disturbances, and the probable multicollinearity of the independent variables. Each of these will be discussed below.

Table 4-4: Summary Statistics for Data Used in the Estimation

Variable	Mean	Standard Deviation	Minimum	Maximum
Innovations (INN_{is})	7.72	24.49	0	365
University Research ($UNIV_{ls}$)	32.52	59.41	0.3	380.60
Industry R&D ($IND_{.s}$)	582.90	818.51	9.0	3883.00
Industry R&D (IND_{ls})	50.87	91.87	0.09	580.80
Related Industry Presence ($RELPRES_{ls}$)	903.06	1062.60	4.04	4404.00
Business Services ($BSERV_{.s}$)	13.88	17.02	0.50	89.52
Geographic Concentration Index ($CONC_{.s}$)	0.41	0.23	1.10	0.94
Industry Sales ($SALES_{i.}$)	9.82	3.48	3.73	16.24
State Population ($POP_{.s}$)	5919.07	4905.33	955.00	22350.00

University research expenditures, industrial R&D expenditures, related industry value-added, receipts from specialized business services and total industry sales are in millions of 1972 dollars. Population is measured in thousands.

The number of innovations in a state and industry is a limited dependent variable which will either be zero or some positive integer. As previously mentioned, there are 95 three-digit industries which contain at least one innovation citation. When the data are stratified by state, a large number of zero cells result. In order to proceed with the model estimation, we limit the sample to the thirteen most innovative three-digit industries listed in Table 4-1. This provides 377 state-industry observations.[7] In the sample used for estimation, there are 140 state and industry observations, or 37%, with zero innovations. Parameter estimates based on only non-zero cases would cause a truncation of the error term which would result in biased estimates. The Tobit Model avoids this problem by using all the cases in the sample to generate more efficient estimates (Tobin 1958). For this reason, we use the Tobit model to estimate the innovation production function equation.

Table 4-5: Correlation Matrix for Innovative Inputs					
	IND_{s}	IND_{ls}	$UNIV_{ls}$	$RELPRES_{ls}$	$BSERV_{s}$
IND_{s}	1.00	--	--	--	--
IND_{ls}	n.a.	1.00	--	--	--
$UNIV_{ls}$	0.68	0.51	1.00	--	--
$RELPRES_{ls}$	0.63	0.58	0.39	1.00	--
$BSERV_{s}$	0.73	0.53	0.56	0.53	1.00

Note: Reported correlations are for the log values of each of the variables. n.a. indicates the variables do not appear in the model together.

Care must be taken in interpreting the estimated coefficients for the Tobit model. The observed number of innovations in a state and industry reflects two underlying processes. The first is whether an area is innovative or not. The second is the number of realized innovations or the measure of exactly how innovative the area is, given that it is innovative. McDonald and Moffitt (1980) demonstrate that a Tobit coefficient can be interpreted as a weighted average of two effects: first, the effect of an increase in an independent variable on the probability that the dependent variable is greater than zero; second, the effect of the expected value of the dependent variable, given that it is above the limit. The magnitude of these two effects depends on the values of the independent variables. As the independent variables reach their maximum, the second effect converges to the value of the Tobit coefficient and the first effect goes to zero.

There are two concerns affecting the possible structure of the error term. The first concern is the spatial auto-correlation of the error terms that is associated with regional cross-sectional data. A random shock affecting economic activity in one state may affect economic activity in adjacent states due to the existence of economic ties between the states. In this case, the disturbances corresponding to contiguous states or states related for other reasons will not be zero and the parameter estimates will suffer a loss of efficiency. Durbin-Watson tests for the existence of auto-correlation were conducted on different orderings of the data. No conclusive presence of auto-correlation was found, therefore no correction for spatial auto-correlation is made. A second concern is the possible heteroscedasticity of the error term. Breusch-Pagan tests revealed no heteroscedasticity in the innovation equation specification.

Table 4-6: Tobit Estimation of Innovation Equation (Standard Errors in Parenthesis)

Variable	Model A:	Model B:
Log($IND_{.s}$)	0.190* (0.054)	
Log(IND_{is})		0.049* (0.034)
Log($UNIV_{is}$)	0.123* (0.044)	0.176* (0.045)
Log($RELPRES_{is}$)	0.296* (0.046)	0.358* (0.050)
Log($BSERV_{.s}$)	0.118* (0.057)	0.212* (0.057)
Log($POP_{.s}$)	0.103* (0.030)	0.131* (0.033)
Log($SALES_{i}$)	-0.230 (0.117)	-0.260 (0.128)
$CONC_{.s}$	1.020* (0.195)	1.113* (0.214)
$\hat{\sigma}$	0.850	0.899
Log-likelihood	-473.46	-477.79

Standard errors are in parenthesis. The Number of observations is equal to 377. Note: * indicates significance of at least .95.

An additional econometric concern in estimating the model is the existence of multicollinearity which is suspect with geographic cross-sectional data. With state data, it is highly likely that the independent variables may be affected by some common trend or underlying state characteristics. Table 4-5 presents the correlation matrix for the innovative inputs. There is evidence of some degree of multicollinearity which may result in parameter estimates exhibiting higher estimated variances and, as a result, the coefficients may be less statistically significant than expected. The next section presents the results of the Tobit estimation.

Empirical Results

Table 4-6 presents the results of the maximum likelihood estimation of the Tobit specification of the innovation equation. Each of the four inputs is statistically significant at 95% for a two-tailed test. Innovative activity in states is related to spill-overs from the innovative inputs of university R&D, industrial R&d, related industry presence and the presence of specialized business services.

Results are presented for the two variations of the industrial R&D variable: Model A uses total industrial R&D expenditures, $IND_{.s}$; Model B uses the estimates of R&D for the related two-digit industry, IND_{is}. This allows an empirical testing of the extent of the technological neighborhood across which R&D spill-overs occur. For example, do R&D spill-overs occur within a narrow band of related industries or do they occur across the entire spectrum of manufacturing industries? A comparison of the two models suggests that Model A, using total industrial R&D expenditures, provides a better model of the innovation equation than the alternative Model B. In so far as any inferences can be drawn, Model A exhibits a better statistical fit. Total industrial R&D expenditures have a larger and more statistically significant coefficient than industrial R&D expenditures estimated at the industry level. The elasticity of innovative output with respect to total industrial R&D expenditures is three times the elasticity of industry specific R&D

expenditures. The higher coefficient on total R&D expenditures may well reflect spill-overs across industries but still within-state boundaries. Furthermore, research and development expenditures are usually undertaken by large integrated corporations with interests in several product categories. The lower coefficient for IND_{Is} may reflect that industry assignment is based on the SIC code of conducting establishment and does not reflect the industry toward which the research expenditures and efforts are directed.

As discussed earlier, the Tobit coefficient estimates cannot be directly interpreted but must be broken down into two effects. At the sample mean of total industrial R&D expenditures, less than one-half of the total response of innovation is attributable to an increase of innovation by states above the threshold; the remainder of R&D expenditures is attributable to an increased probability of achieving some innovative output.

The results in Table 4-6 indicate that not only does innovative output increase in the presence of high private expenditures on R&D, but the presence of research expenditures undertaken by universities within the state also increases innovative output. Industrial R&D is specifically targeted to produce innovative output. The fact that university research expenditures spill-over into increased innovative activity is supported by recent work by Jaffe (1989) and Mansfield (1990). The university research expenditures variable, $UNIV_{Is}$, is measured at the departmental level. The results indicate that the strength of the research expenditures at the academic departments relevant to the innovative industry will have a statistically significant effect on innovative output within the state.[8]

Related industry presence, $RELPRES_{IS}$, is statistically significant as it relates to innovation in a state. This finding indicates that tacit experience with a technology, translates into increased innovative output in a state. Related industry presence is measured as value-added in the larger industry group to which the innovation is attributed. This variable has the largest elasticity of innovative output with respect to an innovative input. This result supports the idea that learning by doing and learning by using are important inputs to the innovation process. The magnitude of this coefficient raises questions about the relationship of industry presence to the other innovative inputs. A strong industry presence in a location may reflect higher industrial and university R&D expenditures in related technical areas. These questions will be considered in the next chapter which presents the estimation of the system of equations designed to account for the endogeneity of the innovative inputs.

The final innovative input is specialized business services. Specialized business services is statistically significant in relationship to innovative output at the state level. This finding is consistent with Alan MacPherson's (1988) finding that localized external sources of specialized business services contribute to realized innovative output.

Three control variables are added to the model: $SALES_{i.}$, $POP_{.s}$, $CONC_{.s}$. Industry sales, $SALES_{i.}$, provides a control for the size of the three-digit industry to which the innovation belongs. As discussed earlier, all of the industries that are used in the analysis are high innovative opportunity industries. This variable provides a control for the total national industry sales which will be related to product demand. This variable was not statistically significant.

Population size, $POP_{.s}$ and the geographic coincidence variable, $CONC_{.s}$, are included to mitigate the effect of aggregation bias caused by using states as the unit of observation. As discussed earlier, states are a less than satisfactory unit of observation. It is difficult to consider that economic spill-overs occur equally in California and in Rhode Island. For this reason, $POP_{.s}$, provides a control for the size of the state and facilitates comparisons across states.

Table 4-7: Distribution of Innovations by Three-digit Industry for Small and Large Firms: Ranked by Number of Total Innovations

SIC	Industry	Number of Innovations		
		Count	Small Firm μ (σ)	Large Firm μ (σ)
357	Computer and Office Machinery	954	18.41 (47.61)	10.52 (21.53)
382	Measuring and Controlling Instruments	668	13.31 (21.64)	7.24 (11.25)
366	Communications Equipment	376	7.66 (15.09)	3.86 (7.58)
367	Electronic Components	261	4.07 (12.53)	4.10 (12.53)
384	Surgical, Medical and Dental Instruments and Supplies	228	5.72 (11.11)	1.59 (3.22)
356	General Industrial Machinery and Equipment	164	3.14 (4.56)	2.24 (2.73)
283	Drugs	133	1.10 (1.97)	3.41 (8.60)
355	Special Industrial Machinery	116	2.45 (3.51)	1.14 (1.85)
349	Miscellaneous Fabricated Metal Products	105	1.83 (3.23)	1.38 (1.99)
362	Electronic Industrial Machinery	74	1.17 (2.02)	1.24 (1.86)
386	Photographic Equipment and Supplies	61	0.97 (1.92)	1.14 (2.29)
282	Plastic Materials and Synthetic Resins	51	0.38 (0.78)	0.76 (1.94)
284	Soap, Detergents and Cleaning Preparations	50	0.83 (1.58)	0.69 (1.79)

The geographic coincidence index, $CONC_{,s}$, provides a control for within-state variation in the concentration of manufacturing activity. One would expect that the more concentrated that the manufacturing activity is within a state, the more likely that the innovative inputs would spill-over into increased innovative output. The geographic coincidence variable is positive and is statistically significant. This result confirms the hypothesis that within-state concentrations of manufacturing activity are important to innovative output.

Small Firm Innovation and Location

For most of the past fifty years, economists have believed that large firm size inherently confers an advantage in innovative activity (Schumpeter 1942; Galbraith 1952). Large, modern corporations benefit from formal R&D laboratories and the internal specialized business services which provide resources to bring new product innovations to market. These resources are typically beyond the means of small business. The research finding that small businesses are the source of innovation in certain industries (Acs and Audretsch 1988, 1990; Pavitt, Robson and Townsend 1987; Rothwell and Zegveld 1982) poses something of a puzzle. Where do small firms acquire the resources to successfully engage in innovation?

One possible explanation is that small firm innovative activity relies on external sources of knowledge as inputs to the innovation process. Larger firms are able to internalize innovative inputs and complementary activities to facilitate innovation. Lacking these resources, small firms may rely on external complementary resources to engage in innovation. The technological infrastructure may enable small firms to engage resources only available to larger firms in other locations. In this way, location may be especially beneficial to small firm innovative activity.

The relative importance of inputs for different firm sizes can be ascertained by estimating the innovation equation separately for the innovative activity of small and large firms and comparing the resulting coefficients. The SBA innovation data identify innovations emanating from large firms and from small enterprises. Large firms are defined as having 500 or more employees. Small firms are defined as having fewer than 500 employees. Table 4-7 presents the distribution of innovation by firm size for the industries used in the estimation.

Table 4-8: Tobit Estimation Results by Firm Size (Standard Errors in Parenthesis)

Variable	All Firms	Large Firms	Small Firms
Log(IND_s)	0.190[a] (0.054)	0.146[a] (0.040)	0.140[a] (0.057)
Log($UNIV_{is}$)	0.123[a] (0.044)	0.068[a] (0.037)	0.145[a] (0.046)
Log($RELPRES_{is}$)	0.296[a] (0.047)	0.245[a] (0.044)	0.214[a] (0.050)
Log($BSERV_s$)	0.118[a] (0.057)	0.071[b] (0.047)	0.149[a] (0.059)
Log(POP_s)	0.103[a] (0.030)	0.085[a] (0.028)	0.126[a] (0.032)
Log($SALES_i$)	-0.230 (0.117)	-0.086 (0.106)	-0.198 (0.120)
$CONC_s$	1.020[a] (0.197	0.810[a] (0.178)	0.895[a] (0.202)
$\hat{\sigma}$	0.850	0.743	0.841
Log-Likelihood	-473.46	-402.24	-444.83

Standard Errors in Parenthesis. Number of observations is 377.
[a] indicates significance at .95. [b] indicates significance at .90.

Table 4-8 compares the results of the model estimation for all firms, large firms and small firms. The four innovation inputs from the geography of innovation model, including university R&D; industrial R&D; and related industry presence are statistically significant at the 95% confidence level. Also, the presence of specialized business services is statistically significant at the 95% confidence level for small firms and at the 90% confidence level for large firms. Clearly, regardless of firm size, the geography of innovation production function for innovative output remains reliable. A Chow test of the equality of coefficients in the two regressions rejects the hypothesis that the two models are equivalent at .01 level of significance. The innovative activity of small and large firms reflect different usage of the innovative inputs.

There are several differences between small and large firms in the importance of the inputs in generating innovative output. Innovative small firms make greater use of university research than do their larger counterparts.[9] The elasticity of innovative output with respect to university research is more than twice as great for small firms than for large firms. One potential explanation for this finding is that small firms have a less well-developed internal R&D capability than their larger counterparts and thus rely more heavily on external R&D from universities. Link and Rees (1990) reach a similar conclusion in survey work on the usage of university research. The lack of internal resources to support in-house specialized business services may also account for the finding that innovative activity of small firms benefits more from the presence of specialized business services in the state.[10] The elasticity of innovative output with respect to business services is twice as great for small firms than for their larger counterparts.

Where do small firms acquire the resources to successfully engage in innovation? The results suggest that small firms rely more on external sources of input to the innovation process. Small firms appear able to generate innovative output while undertaking negligible amounts of investment in R&D by capturing spill-overs from university research. Large firms are more adept at exploiting knowledge created in their own laboratories, while smaller firms exhibit a comparative advantage at exploiting spill-overs from university laboratories. One possible explanation for this is that small businesses have a less well-developed internal R&D capability than their larger counterparts and thus rely more heavily on external R&D from universities. While large firms are more likely to have formal university research relationships, small firms tend to receive greater benefit. Another possible explanation of this result is that knowledge which may be otherwise difficult to appropriate results in the start-up of a new, small firm. Further, the findings indicate that a small firm's innovative activity benefits more from the presence of specialized business services than their larger counterparts. A plausible explanation for this finding may be the small firms' lack of internal resources to allocate to in-house, specialized business services. This finding is consistent with MacPherson's (1988) findings that the intensity of use of external producer services correlate highly with small firm product innovations.

Larger firms are able to internalize innovative inputs and provide complementary activities that facilitate innovation. Lacking these resources, small firm innovative activity appears to benefit from an external technological infrastructure, an integrated and spatially-concentrated network of institutions which provide inputs to the innovation process. The technological infrastructure may enable small firms to engage resources that were once only available to larger firms in other locations. In this way, location may be especially beneficial to small firm innovation.

In conclusion, the main findings of the empirical estimation confirm the hypothesis that innovation is concentrated in places that possess a well-developed technological infrastructure. Innovations are found to cluster geographically in areas that contain concentrations of specialized resources which comprise a technological infrastructure. These findings support the conceptualization that concentrations of knowledge are important to new product commercialization. The next chapter explores the relationship between these knowledge resources.

Chapter Notes:

1. The geographic concentration variable, $CONC_{,s}$, measures the amount of manufacturing activity in the largest Standard Metropolitan Statistical Area (SMSA) relative to total manufacturing activity in the state. These data are presented on page 127 of the Appendix. For the estimation of the innovation equation, the log of the geographic concentration variable is not taken. There is no strong *a priori* functional specification and the estimation of the innovation equation with a log transformation of this variable yielded similar results.

2. The dependent variable is based on the SBA innovation citation data introduced in Chapter 1 and discussed in Chapter 3.

3. A recent survey by Edwin Mansfield (1991) estimates that the average time lag between a recent academic research finding and the first commercial introduction of a new product was on average approximately 7 years with a standard deviation of 2 years. These results are based on a survey of key managerial technical personnel from 76 large corporations in 7 industries.

4. The data were compiled by the Neil Bania at the Center for Regional Economic Issues, Case Western University.

5. The sample for estimation accounts for 78% of the U.S. population and 81% of the university research expenditures in 1982.

6. An alternative to estimate key suppliers and users would be to use the 1978 National Input-Output Tables to link three-digit industries together (Survey of Current Business). This approach revealed that the most important linkages occurred within the larger two-digit industry. For example, the industry SIC 283, Drugs, received an input share of 27% from related industries in SIC 28. This was the single largest share with the remainder of 73% of inputs dispersed across other industries.

7. Three hundred and seventy-seven observations are the result of data for 13 industries and 29 states.

8. An alternative specification of the innovation equation using total university research expenditures, similar to the specification presented using two variations of industrial R&D, was estimated. Total state university research expenditures were statistically significantly related to innovations within the state, however the overall fit of the model declined. Pragmatically, this is understandable because it is less plausible that basic research in biology would spill-over to innovations in an unrelated field such as electronics.

9. There is a statistically discernable difference in the coefficients of university research expenditures for small and large firms at the 95 percent level of confidence using a two-tailed test.

10. There is a statistically discernable difference in the coefficients of business services for small and large firms at the 90 percent level of confidence using a two-tailed test.

5
Regional Innovative Capacity

Case studies of successful high-technology complexes such as Route 128 (Dorfman 1983) or Silicon Valley (Saxanien 1985) suggest that these areas and industries developed as entrepreneurial efforts built on the existing technological infrastructure. While historical events and specific individuals served as catalysts in the development of these areas, these regions benefitted from local institutions and resources which provided a core competency to move both the area and the industry forward. This technological infrastructure creates an underlying capacity that supports and sustains innovative activity. Part of the classical story of innovative locations is the mutually dependent and reinforcing nature of these different types of knowledge.

Our results demonstrate that commercial innovation benefits from spill-overs from the technological infrastructure. This technological infrastructure includes the formal technical knowledge that is derived from university research and industrial R&D, the tacit knowledge which accompanies familiarity with a technology, and the commercial knowledge of the market. The location of these innovative inputs are not exogenous to the model of the geography of innovation. Relationships governing the innovation inputs are important to understanding how areas develop a comparative advantage for innovative activity in specific industries. The mutually-reinforcing and path-dependent nature of knowledge creates the specialized capabilities which propels innovation forward in particular technologies and industrial sectors. Once in place, this technological infrastructure nurtures opportunities for increased innovative activity. This chapter considers the determinants of the location of the resources in the technological infrastructure.

The Capacity Building Perspective

Classical location theory focuses on the individual location decisions of profit-maximizing firms. According to this perspective, firms freely scan the landscape to select locations which optimize the firm's functional requirements.[1] In studying innovation, this perspective is especially limited because it focuses on the more traditional factors of production and does not consider knowledge inputs. In contrast to information, knowledge is the result of a stock of significant investments made over time. Knowledge will be embodied in either institutions or in human beings and it is expected to be relatively immobile and place-specific. Knowledge is created through social relationships and in specific contexts. It is these knowledge creating relationships which provide the capacity to commercialize new products. In this sense, it is the regional capacity which creates the potential to innovate and which shapes, and is shaped by, the locational choices of individual firms.

Some scholars argue that places grow partly as a result of serendipity or so-called "historical accidents"(Storper and Scott 1990; Arthur 1900). While there is no doubt that such unpredictable chance events occur, these "accidents" will not spark innovation and subsequent economic development if an area lacks a technological infrastructure. The ability of an area to capture the benefits of historical serendipity is determined by the level of preparedness of the underlying technological infrastructure. Chance occurrences may not materialize into any enduring effect if key knowledge resources are lacking. Indeed, scientific discoveries may be developed and commercialized elsewhere and new firm starts-ups may migrate to more fertile areas if the knowledge that promotes these innovative activities is lacking. Locational advantage and innovative capacity appear to stem from, and to be embedded in, the underlying technological infrastructure of a place.

There is an emerging literature that considers the role of regional capacity, defined here as the ability of place-specific resources to promote economic activity. Florida and Kenney (1988) suggest that innovative activity is promoted by an underlying social structure of innovation which promotes and sustains new firm start-ups. Storper and Walker (1989) emphasize the spatial and place-specific nature of the process of technological change and industrial development. We have, however, a limited understanding of the nature of place-specific advantage and how it comes into being. The question at hand is -- what determines a regional's innovative capacity?

Empirical Model

A region's innovative capacity is modeled as a recursive system of four equations. Individual equations are specified to isolate the determinants of industrial R&D, university R&D, and business services. We treat the presence of firms in related manufacturing industries as exogenous.[2]

Innovative output, INN_{is}, the number of innovations for an industry, i, and a geographic area, s, is modeled as a function of four innovative inputs:

$$INN_{Is} = UNIV_{Is}^{B_1} \, IND_{Is}^{B_2} \, RELPRES_{Is}^{B_3} \, BSERV_{Is}^{B_4}. \quad (8)$$

IND_{Is} is industrial R&D expenditures in the geographic area, s, for the related, larger industry group, I. The more encompassing industry group, *I*, measures the technological area across which spill-overs would be expected to occur. $UNIV_{Is}$ is university research expenditures measured at the level of the academic department and allocated to the relevant industry group. $RELPRES_{Is}$ is the presence of related industry, including the presence of firms in industries using related technologies as well as downstream users of a technology who may disseminate information about the technology. $BSERV_{Is}$ is the presence of specialized business services related to innovative activity within the industry.

Equation (8) approximates a technical relationship that relates the location of innovative inputs to the location of realized innovative output. The presence of the innovative inputs in a geographic area will be a function of a set of behavioral relationships which govern their location.

The second equation in the system examines the factors that determine the location of industrial R&D. Malecki (1985) found that R&D laboratories tend to locate either near production facilities or near firm headquarters. These locational patterns reflect the historical development of industrial R&D (Mowery and Rosenberg 1988). We also expect that industrial R&D expenditures would be allocated to regions where there are strong related university research programs. Mansfield (1991) found that nearly 40 percent of the researchers cited by electronics and information processing firms and a quarter of the researchers listed by chemical and pharmaceutical companies were located in the same state as the firm that cited the work as important to new product development. Jaffe (1989) found a strong relationship at the state level between industry and university R&D. The industrial R&D equation examines the location of industrial R&D in relation to the location of other innovative inputs, notably university research, firms in related industries, and corporate headquarters.[3]

The industrial R&D equation is specified as:

$$\log(IND_{s}) = \gamma_1 \log(UNIV_{s}) + \gamma_2 \log(RELPRES_{s}) + \gamma_3 \log(HDQRT_{s}) + \gamma_4 lg(POP_{s}) + \varepsilon_{Is}. \quad (9)$$

where $RELPRES_{.s}$ is related industry presence in the area, and $HDQRT_{.s}$ is included to measure the presence of Fortune 500 manufacturing firm headquarters in the area.

The third equation in the system examines the factors which effect the location of university R&D. As noted above, Mansfield (1991) suggests that industry R&D expenditures and presence will influence university research expenditures in the geographic area. Mansfield found that companies tend to use research at nearby universities, independent of the generally perceived quality of the school. Companies tend to be more aware of research projects that are being conducted at universities within their same state and are more likely to use this research in their product development.[4] As a result, firms may be more likely to fund research at these universities or otherwise contribute to the ability of the related academic department to attract funding. Jaffe (1989) provides strong evidence of the co-location of industrial and university R&D.

The university R&D equation explores the interaction between university research, industrial R&D and other determinants. It is specified as:

$$\log(UNIV_{IS}) = \gamma_1\log(\Sigma IND_{Is}) + \gamma_2\log(\Sigma RELPRES_{Is}) + \gamma_3 FFRDC_{.s} + \gamma_4\log(POP_{.s}) + \upsilon_{IS}. \quad (10)$$

where $UNIV_{Is}$ represents research expenditures for departments with research relevant to an industry group, I.; the summation operator on industrial R&D expenditures, IND_{Is} and value added, and $RELPRES_{Is}$, indicates the total for these variables across industries for whom the academic department's research is relevant. Table 5-1, which is a re-representation of Table 4-2, links university departments back to the two-digit industrial R&D expenditures and the related industry presence considered pertinent to the university department. For example, research in the chemistry and chemical engineering departments is relevant to Chemicals (SIC 28). $FFRDC_{.s}$ is the number of Federally-Funded

Research and Development Centers in each state and serves as a barometer of a university's receptiveness to participate in technology transfer with industry. This equation links university research expenditures in specific academic departments to the presence of related industrial R&D expenditures and related industry in the state.

Table 5-1: Linking University Departments to Relevant Industries

University Department	Industry
Chemistry and Chemical Engineering	SIC 28: Chemicals
Medicine and Biology	SIC 28: Chemicals SIC 38: Instruments
Mechanical Engineering, other engineering and physical sciences	SIC 34: Fabricated Metal Products SIC 35: Industrial Machinery SIC 38: Instruments
Electrical Engineering, Astronomy, Physics, and Computer Science	SIC 35: Industrial Machinery SIC 36: Electronic Equipment SIC 38: Instruments

The fourth and final equation in the system examines the factors which effect the location of business services. More specifically, this business services equation examines the location of specialized business services producers tailored to innovative activity. The presence of specialized producer services is expected to reflect the client base from which these services draw their existence, such as industrial R&D laboratories. This innovative input provides the closest link to market introduction. However, no simultaneity with current innovative output is modeled. The equation is specified as follows:

$$\log(BSERV_s) = \alpha_1 \log(IND_s) + \alpha_2 \log(TOTALBSERV_s) + \alpha_3 \log(POP_s) + \xi_s. \quad (11)$$

where $TOTALBSERV_s$ is general producer services.

Table 5-2: Summary Statistics for Data Used in the Estimation

Variable	Mean	Standard Deviation	Minimum	Maximu m
Innovations (INN_{is})	7.72	24.49	0	365
University Research ($UNIV_{Is}$)	32.52	59.41	0.3	380.60
Industry R&D ($IND_{\cdot s}$)	582.90	818.51	9.0	3883.00
Industry R&D (IND_{Is})	50.87	91.87	0.09	580.80
Related Industry Presence ($RELPRES_{Is}$)	903.06	1062.60	4.04	4404.00
Business Services ($BSERV_{\cdot s}$)	13.88	17.02	0.50	89.52
Geographic Concentration Index ($CONC_{\cdot s}$)	0.41	0.23	1.10	0.94
Industry Sales ($SALES_{i}$)	9.82	3.48	3.73	16.24
Sales of Fortune 500 Firms ($HDQRT_{\cdot s}$)	30,281	55,249	100.00	271,700
FFRDC ($FFRDC_{\cdot s}$)	0.59	1.00	0.00	4.00
General Business Services ($TOTALBSERV_{\cdot s}$)	443.21	545.71	40.71	2440.00
ΣIND_{Is}	84.49	120.35	0.49	603.90
ΣVA_{Is}	1732.10	1901.10	59.17	8381.00
State Population ($POP_{\cdot s}$)	5919.07	4905.33	955.00	22350.00

Innovations is measured as integer count of innovations. University research expenditures, industrial R&D expenditures, related industry value-added, receipts from specialized business services and total industry sales are in millions of 1972 dollars. Population is measured in thousands.

In conclusion, the model developed here provides a means for empirical investigation of the spatial relationships that are associated with innovation. This model relates innovation output to the location of innovative inputs and explores the interrelationships between the knowledge inputs and the commercialization process. A summary of the data used in the empirical estimation is presented in Table 5-2. Innovations are measured as the count of innovation in a state and three-digit industry. University research expenditures, industrial R&D expenditures, related industry value-added, receipts from specialized business services were described in Chapter 4. The commercialization process is characterized by a time-lag from invention to actual market introduction. While this time-lag is variable and difficult to specify, a recent study by Mansfield (1991) estimates the average time-lag between academic research finding and commercial introduction of a new product is seven years with a standard deviation of two years. Taking this into account, we use the average annual expenditures for the variables in the ten-year period prior to the introduction of the innovation. Sales for Fortune 500 Firms is from *Fortune Magazine*. Company sales are allocated to the state in which the firm's headquarters are located. State population is measured as the resident population, in thousands, in 1977. Industry sales are measured as the 1977 value of shipments for the three-digit industry. To reiterate, the year 1977 is used to estimate the time-lag inherent in the commercialization process.

The four equations are estimated as a recursive system. This allows us to empirically test the ways in which the technological infrastructure affects the innovation process, and to isolate the factors that act on each of the components of that infrastructure. The system is recursive in the sense that there is no direct feedback from the second, third and fourth equations to the first equation. Also, the system is not simultaneous in the usual sense because of the expected time-lag involved in translating successful innovative output into increased expenditures on innovative inputs.

The interrelationships between the variables indicate that the efficiency of the parameter estimates will increase with simultaneous equation estimation. The four equations are estimated together using Three Stage Least Squares (3SLS) Regression. The potential of correlations across the equations dictates the use of instrumental variables.

The instruments include all of the exogenous variables appearing on the right hand side of the equations in the model.

There are some statistical issues to consider in the model estimation. These include the censored nature of the dependent variable and the likely structure of the error terms. The dependent variable, the number of innovations by state and industry, is a censored dependent variable. The number of innovations will either be zero or some positive integer. Cases with no innovations provide information about how innovative locations differ from non-innovative locations. Customary estimation of a production function such as equation (8) relies on a log-log transformation. The zero observations present a problem in this regard. To estimate the innovation equation, the dependent variable, INN_{is}, is transformed to eliminate zero values. The new dependent variable is Log(10*(1+INN_{is})). This transformation preserves the relative ranking of the innovation observations.

There are two potential concerns regarding the error term. The first pertains to spatial auto-correlation and its association to regional cross-sectional data.[5] Durbin-Watson tests for the existence of auto-correlation were conducted on different orderings of the data and no conclusive presence of auto-correlation was found. Therefore, no correction for spatial auto-correlation is made in the estimation. The second concern is the possible heteroscedasticity of the error term, however no evidence of heteroscedasticity in the innovation equation specification was found.

EMPIRICAL RESULTS

The results of the estimation of the model are presented in Table 5-3. Generally speaking, the model performed well. The coefficients of the four innovative inputs, industrial R&D, university research expenditures, related industries, and business services, that represent

Table 5-3: Estimation Results on the System of Equations	
Dependent Variable: Log(10*(INN_{is} + 1))	
Log(IND_s)	0.241[a] (0.054)
Log($UNIV_{is}$)	0.155[a] (0.043)
Log($RELPRES_{is}$)	0.144[a] (0.045)
Log($BSERV_s$)	0.272[a] (0.055)
Log(POP_s)	0.054[b] (0.030)
Log($SALES_i$)	-0.236[a] (0.113)
$CONC_s$	1.021[a] (0.189)
Dependent Variable: Log(IND_s)	
Log($UNIV_s$)	0.566[a] (0.074)
Log($RELPRES_s$)	0.466[a] (0.089)
Log($HDQRT_s$)	0.180[a] (0.040)
Log(POP_s)	0.0486[b] (0.026)
Dependent Variable: Log($UNIV_{is}$)	
Log(ΣIND_{is})	0.256[a] (0.039)
Log($\Sigma RELPRES_{is}$)	0.338[a] (0.047)
$FFRDC_s$	0.539[a] (0.054
Log(POP_s)	-0.118[a] (0.031)
Dependent Variable: Log($BSERV_s$)	
Log(IND_s)	0.109[a] (0.040)
Log($TOTALBSERV_s$)	0.707[a] (0.057)
Log(POP_s)	-0.295[a] (0.215)

Note: Standard errors are in parenthesis. The instruments used include all of the exogenous variables appearing on the right-hand side of the equations in the model. The number of observations is equal to 377. [a] indicates significance of at least .95. [b] indicates significance at .90.

the technological infrastructure, are all positive and statistically significant. A number of permutations of the model were run to ensure that the findings were robust. The model was run with each variable measured on a per-capita basis. The basic results, signs and significance for the coefficients remained the same. The model was also run with geographic-fixed effects to test the robustness of the results for states with more than one large manufacturing center and for contiguous states. Here again, the basic results did not change. Therefore, we conclude that the results are robust. Simply put, the empirical results confirm that innovative activity within states is related to the underlying technological infrastructure.

Looking first at the innovation equation, the coefficient estimate for industrial R&D is positive and significant. The size of the coefficient further suggests that industrial R&D plays a strong positive role in the innovation process. This is generally in line with the findings of the extant literature, that industrial R&D plays an important role in the innovation process. University R&D similarly has a positive and significant coefficient. This indicates that university R&D plays an important role in the innovative capacity and technological infrastructure of a given place.

The coefficient estimate for the presence of related manufacturing industries is positive and significant. Concentrations of manufacturing firms in related industries have a positive effect on an area's capacity to innovate. This supports the notion that networks of related firms and their suppliers provide important knowledge inputs to the innovation process. Synergies between and among firms in related industries appear to play an important role in the innovation process. The size of the coefficient for business services indicates that they also have a particularly positive effect on the innovation process.

Turning now to the sub-equations in the model, the results for the industrial R&D equation indicate that the location of industrial R&D

expenditure is positively related to university R&D expenditures, concentrations of firms in related manufacturing industries, and the presence of large firm headquarters. The coefficients for each of these variables are positive and statistically significant. The magnitude of the effect associated with university R&D expenditures suggests the existence of a strong relationship between university research and industrial R&D. Again, this supports the earlier research findings (Jaffe 1989). Further, the size of the coefficient for related industry is suggestive of a rather strong relationship between industrial R&D and the broader industrial base. This is not surprising as industrial R&D tends to both feed off of and support local manufacturing activity.

The empirical findings for the university R&D equation indicate strong relationships between university R&D, industrial R&D, and the presence of related industries. There is a somewhat weaker, though still significant, relationship between university R&D and the presence of corporate headquarters. University R&D expenditures are positively associated with both industrial R&D and the presence of firms in related industries. The coefficient for related industry presence is highly and statistically significant with regard to university research expenditures at the department level. This result suggests that university research provides a source of information for technology that is relevant to industrial activity within the state, or alternatively, that expenditures for university research reflect the state industrial base. The coefficient for federally-funded research centers is positive and statistically significant. Furthermore, the results suggest that the expenditures for university R&D at the department level are associated with industrial activity and industrial R&D in related fields and industries. This is in line with Mansfield's (1991) finding that firms tend to use research from nearby universities. Overall, the results here point to a significant degree of regional specialization in R&D activity.

In comparing the results of the industrial R&D and the university R&D equations, it appears that the effect of university R&D on industry R&D is greater than vice-versa. This in turn suggests that university R&D may play a particularly important role in the innovation process, by attracting industrial R&D, or at least by very positively leveraging industrial activities. However, the relatively smaller effect of industrial R&D on university R&D may be explained by the fact that a large proportion of total university R&D is provided by the federal government. In 1982, almost two-thirds of total university research funds came from federal sources.

The findings of the business services equation indicate that the presence of specialized business services are positively related to industrial R&D and the size of the general business services sector. This suggests that specialized business services are located in the same areas where industrial R&D laboratories, their expected clients, are located. In addition, concentrations of specialized producer services appear to be related to more general concentrations of business service activity.

Taken together, these findings provide a greater understanding of the dynamics of an area's technological infrastructure. First, they are suggestive of the considerable regional specialization of industrial activity and university-based R&D. Second, they suggest a considerable level of mutual reinforcement and synergy among the four major components of that technological infrastructure. The location of industrial R&D is associated with the location of university R&D, and the presence of related industries. University R&D is associated with industrial R&D, the presence of related industries, and federally-funded research centers. Furthermore, industrial R&D, university R&D, the presence of related industries, and business services are all positively associated with the capacity of an area to innovate. Thus, the four institutional factors work together in a synergistic way in specific places to comprise an interactive infrastructure that supports innovation. The innovative capacity of a place -- in this case a state -- turns on this underlying technological infrastructure. The empirical evidence confirms the hypothesis that geography plays a significant and important role in organizing and mobilizing the innovation process.

The findings of the model confirm the hypothesis that innovation is a function of an area's technical infrastructure. Innovation is positively related to geographic concentrations of industrial R&D, university R&D, related industries and business services. The recursive system of equations suggests that there is significant synergy and mutual reinforcement among the innovation factors. The empirical findings further indicate that there is considerable geographic specialization in the technological infrastructures of various places. Clearly, the innovative capabilities of particular places stem from specialized concentrations of specific fields of university R&D, industrial R&D, and the presence of closely related industries.

The innovative capacity of an area may be seen as the result of historical events that determine a growth trajectory or path of a geographic area (Arthur 1990). Core competencies emerges as resources become tailored to activity within an industry. The co-location of complementary resources provide economies of scope which benefit innovation and new product commercialization. Taken together, the complementary institutions of the technological infrastructure provide resources and knowledge inputs to the innovation process, generate positive externalities and spill-overs which lower the cost of commercializing new products, and reduce the risks associated with innovation. The ability to bring new product innovations to market is demonstrated to increase as a result of the technological infrastructure. The next chapter examines the policy implications of our findings.

Chapter Notes:

1. The limitations of this conceptual framework have been considered elsewhere. For example, see Clark, Gertler, Whitman (1986).

2. The location of the related-industry presence associated with innovation may be best considered as part of a temporal model of technological change. An area that develops an endowment of resources tailored to foster innovative activity in an industry will provide a strong locational advantage for firms in the industry over other areas. The importance of linkages among related firms that draw an industry to an area is a very provocative idea, however there is little empirical evidence upon which to build. Sveikauskas, Gowdy and Funk (1988) suggest that proximity to suppliers and similar firms is more important to industry productivity than scale economies, based on a study of the food processing industry. Justman (1990) finds important linkages between local demand and industrial growth, concluding that industries which rely on information externalities, such as semiconductors and metalwork machinery, exhibit stronger market linkages than those that do not.

3. No simultaneity between an area's innovative success and industrial R&D allocation is estimated because of the time-lag in introducing an innovation to the market. Industrial R&D is typically 4-5 years removed from market introduction and any firm's response to successful innovative output will reflect that time delay.

4. Mansfield (1991) found that nearly 40% of the researchers cited by electronics and information-processing firms and 25% of the researchers listed by chemical and pharmaceutical companies were located in the same state as the firm that cited the work as important to new product development.

5. A random shock affecting economic activity in one state may affect economic activity in adjacent states due to the existence of economic ties between the states. In this case, the disturbances corresponding to contiguous states, or states related for other reason, will not be zero and the parameter estimates will suffer a loss of efficiency.

6
Innovation Policy

Geography clearly plays a role in facilitating innovation and in furthering new product commercialization. Indeed, our findings suggest that innovation is in itself a geographic process -- a function of the knowledge resources that are embodied in the technological infrastructure of specific places. A technological infrastructure provides locational advantage for innovation, promotes technological advance in an industry and may further the economic fortune of regions. This locational advantage, rather than the result of endowments of natural resources, or transportation-cost differentials, or the availability of lower-cost labor, stems from the presence of complementary and mutually reinforcing knowledge resources. Our results indicate that regions with a strong knowledge-based technological infrastructure realize greater numbers of product innovation.[1]

This Chapter considers the implications of our findings for innovation policy. We consider the perspective of private companies which are engaged in new product commercialization. Our results suggest that the importance of geographically-mediated spill-overs provide a broader criteria for firms' location and investment decisions. In addition, we consider the perspective of state governments which are trying to promote economic development. The public good nature of knowledge suggests a role for state government in building and augmenting the knowledge resources that contribute to the commercialization process.

Implications for Companies

By many indicators, U.S. research capabilities are preeminent in the world. However, this lead appears to be lost during the translation of scientific knowledge into new commercial-product innovations. Unfortunately, American firms suffer from longer new-product development cycles and there is evidence that a wide variety of U.S. industries have failed to commercialize new innovations effectively and rapidly.[2] This evidence, coupled with America's declining trade performance, has provoked wide debate over the need to find new ways to organize innovative activity. While this study has not focused on the practices and policies of individual companies, the results suggest that product commercialization benefits from location.[3]

Consider the following quote from Robert Noyce, founder of the Intel Corporation.

> "Our Industry tends to cluster geographically. Why? Because it is to take advantage of the infrastructure of talent pools, support services, venture capital and suppliers. Our industry tends to use the same suppliers of equipment, spreading the development cost of the equipment broadly. We tend to use the same software in the design, manufacturing, financial or distribution systems, reducing the cost of acquisition for each of us. We also serve the same customer in most cases, resulting in defacto standards on pin configuration, functional specifications, packaging, operating voltages and the like. [These activities are a form of localized voluntary cooperation.] The other type of cooperation we might call involuntary. That is the mobility of our personnel which quickly diffuses knowledge of new techniques in design, production, and marketing throughout the industry. ... Stop for a moment and visualize the structure of our industry if [information access had been limited.]....We would be in the Dark ages in comparison to the present state of the industry.....We would be trying to understand the cacophony of the Tower of Babel instead of talking about the microprocessor revolution."

In this quote, Robert Noyce (1982: 14) discusses the benefits which location can provide to innovative activity. Noyce's contemporary assessment mentions the same factors noted by Alfred Marshall many years earlier. The critical element for innovative industries and companies appears to be the knowledge externalities that facilitate innovation and new product commercialization. While, the spill-over mechanism may be voluntary or involuntary, the critical element appears to be location. The computer industry, brought to the forefront by explosive growth, provides a well-known, contemporary example of the potential benefits of location. However, the geographic concentration of product innovation within other industries suggests the existence of a similar process.

All policy decisions are predicated on models, or abstractions, of reality, and company policy regarding innovation is no exception. The oldest model of the commercialization process, the linear model, places great importance on R&D capacity, especially in-house R&D (See Figure 2-2). Indeed, in the linear model, R&D and scientific discovery are the sole source of new product ideas. This discounts other types of knowledge and expertise. However, when we re-conceptualize the commercialization process to include feedback and linkages among the various stages, then different types of knowledge become sources of innovation. The knowledge resources may exist within an organization or be external to the innovating firm. Specifically, the practical types of knowledge associated with manufacturing, and with marketing and business services may not be considered important. Dertouzos et al. (1989) argue that companies and universities need to change their attitudes towards other types of knowledge. In many instances, the tacit knowledge gained from working with a production process is viewed as an activity of lesser importance.

Our conceptualization of the commercialization process suggests that linkages between the sources of knowledge are critical. This argues favorably for increased interaction between the knowledge sources. Each of the four innovative inputs contains a critical stock of knowledge for the commercialization process. When all of the types of knowledge are present, the process is more likely to be successful. This is part of the rationale behind inter-departmental project teams in new product development. Our results suggest that co-location of the sources of knowledge in the same geographic area may increase idea exchange and communication. Product commercialization will be more successful if knowledge synthesis occurs easily and at lower cost. To accomplish this our results suggest that co-location or geographic-consolidation of new product development efforts may be beneficial.

Also, our findings highlight the complementarities between the knowledge sources. This includes activities which occur within the boundaries of the firm and also extends to external sources of knowledge. We find that a proximity to external sources of knowledge is related to higher levels of innovation. Firms will be more innovative if they are in proximity to the sources of knowledge for commercialization and cultivate external linkages to suppliers, product-users and firms using similar technology. The importance of external knowledge sources suggests that companies' location decisions reflect the opportunity to acquire critical knowledge in a timely manner and to easily tap into existing sources of commercialization knowledge. Most critically, organizations may need to be in locations which provide the necessary sources of knowledge. This appears to be especially important for small firms that are less able to internalize the expertise for new product commercialization.

An emphasis on the importance of tacit knowledge is also pivotal in company location decision-making. The definition of competition is changing, shifting away from an emphasis on price competition to one that includes quality, product variability and market response time. Our results argue that R&D expenditures may be more productive where there is a strong related-industry presence. Florida and Kenney (1990) have argued that the geographic separation of design, manufacturing and assembly undermines an industry's ability to improve products or to respond quickly to market changes. The results presented here argue that the economies of scope provided by resources used in the commercialization process may be an important factor for company location decisions.

While the role that geography plays is an important one, companies are not passive actors in this process. Companies play a role in defining the innovative capacity of regions. A firm's investments in the technological infrastructure will yield benefits for the company and for the region. Our findings argue for a fuller understanding of the importance of building knowledge assets. Developing knowledge capabilities in an organization requires long-term planning horizons. It is difficult to gauge the contribution of knowledge-building activities to profitability. Private firms' lack of incentive to engage in building knowledge-based resources provides a rationale for state government involvement.

The Role of State Government

States are increasingly attentive to issues of fostering innovation and innovative behavior as a means of promoting economic development (Osborne, 1988; Schmandt and Wilson, 1990). A geographic cluster of product innovation creates a potential locus of economic growth as firms expand to meet new product demand. Theoretical models indicate that the geographic clustering of innovative activity is due to increasing returns to non-transferable activities (Arthur 1990; Grossman and Helpman 1989). This suggests that state efforts to build an innovative infrastructure are not transferable to other locations. As a result, states can expect to capture a return on their infrastructure building efforts.

Conservative policy-makers argue that the market mechanism will take care of the innovation process and government intervention is not warranted (Gilder 1988). However, this view fails to recognize the public good nature of knowledge-based inputs to the innovation process. Scientific, technical and commercial knowledge provide public intermediate goods which simultaneously benefit groups of related firms. The non-exclusionary nature of these resources suggests that private firms have an incentive to under-invest in their provision. State policies that encourage more efficient use of these resources will result in greater efficiency for the entire economy.

This is not to imply that states can control or define the innovation process. State efforts operate, to a large extent, within confines that are set by the Federal government. For example, national policy determines the single largest source of university research funding; Federal tax policy determines incentives for industrial R&D; and, Federal trade policy determines the size of the market and the competition which new products and new industries face.

Other aspects of innovative activity are inherently local in nature (Saxenian 1991). The complexity of the social interactions involved in the innovation process suggests that local institutions need to be given autonomy and operational flexibility. Perhaps, the role of the state may be to provide resources to allow local entities to capitalize and to build on existing strengths. At a minimum, states should not co-opt or disrupt local success. Ultimately, the major actors in the innovation process are firms and firms respond to market signals and opportunities. Individual firms decide which product categories to target, how many resources to allocate to innovation and how to search for knowledge to use in commercializing new products. What state governments may do is act as umbrella organizations in supporting the knowledge-based innovative inputs which lower the costs of innovation to private firms.

A fundamental stumbling block for state economic development has been a focus on employment impacts. It is easy to understand why jobs are an important concern for policy-makers. However, employment gains may not be very useful criteria for long-term economic development purposes. Areas that nurture high-technology industrial employment will eventually find that these industries share in the global mobility of standardized production and attraction to low wages that have become so familiar with more traditional industries (Bluestone and Harrison 1982). In addition, an emphasis on an employment-based definition focuses efforts toward attracting high-technology branch plants which increase area employment. In many instances, however, these jobs are typically low-paying and the positions are not secure. In time, these jobs easily shift to lower-cost locations (Cobb 1982).

Thompson (1968) argues, in contrast, that the high-technology economic base of a region should be defined by the innovative activity located there. If a region is able to continuously generate and promote innovation, then it will be able to adapt to changing economic circumstances. The capacity to innovate will determine the regions's ability to generate new products, and new firms and industries as economic and technological conditions change. Malecki (1991) suggests that in an effort to adapt to structural economic change, regions promote non-routine production activities such as research and development, experimental and prototype manufacturing, and the small volume production of new, changing products. These are the activities that are supported by the technological infrastructure.

Many state economic development strategies are targeted to a narrow definition of high-technology industries and ignore the potential of existing industries. Feller (1984) criticizes the emphasis of many state programs that are targeted to biotechnology, microelectronics or robotics as "high-tech hype." These programs focus resources on the more glamorous sectors and neglect the existing industries in which the state may have already developed a strength or comparative advantage. In addition, our results indicate the importance of geographic clustering. All places cannot simultaneously be the site of a similar agglomeration. Indeed, competition among states may dissipate the resources needed to provide the critical mass to support the success of an emerging industry.

Table 6-1 presents the distribution of product innovations among manufacturing industries. Many of the industries that account for a large number of innovations are not considered high-technology by widely-used definitions (e.g. U.S. Office of Technology Assessment 1984). Because these industries lack that glamorous title, they may be overlooked by state economic development efforts. To avoid this myopia, programs like the Ben Franklin Partnership focus their attention on the diffusion of advanced technology to companies in the state. The idea behind the Ben Franklin Partnership is to channel scientific knowledge to revitalize and increase the capacity of existing industries.

Traditionally, in the United States federalist system, states are assigned the tasks of overseeing knowledge-based resources and of building infrastructure. Our findings suggest that attention to this traditional role offers states a means to promote economic development. This should not be interpreted to suggest that state governments are able to define and control the commercialization process. While innovative activity is no longer confined to the organizational boundaries of the firm, we should not underestimate the importance of firms and entrepreneurs in the commercialization process. The actions of entrepreneurs, in turn, are driven by technological opportunities and market forces. At best, states are an intermediary with a limited but well-defined role. States can strategically allocate their resources to build on existing strength and expertise within their boundaries, and thereby promote innovative activity and engender endogenous growth. Although the process entails a lengthy and consistent commitment, it offers states a path to obtain a sustained economic advantage.

Table 6-1: State Comparative Advantage in Innovative Industries

Industry	N	Leading State	n	Location Quotient
Computers	954	California	356	167.8
Measuring Instruments	668	California	134	126.4
Communications Equipment	376	California	116	132.2
Electronic Equipment	261	California	128	211.3
Medical Instruments and Supplies	228	New Jersey	57	248.2
General Industrial Machinery	164	Pennsylvania	25	261.5
Drugs	133	New Jersey	52	381.3
Special Industrial Machinery	116	Illinois	11	171.4
Misc. Fabricated Metal Products	105	Ohio	18	384.0
Electronic Industrial Machinery	74	California	17	94.4
Photographic Equipment	61	New York	18	260.0
Plastic and Synthetic Materials	51	Texas	10	491.7
Cleaning Preparations	50	New York	10	183.3
Refrigeration Machinery	49	Wisconsin	6	583.3
Misc. Plastic Products	47	New York	6	118.2
Office Furniture	47	Illinois	8	318.2
Construction Equipment	43	Ohio	8	430.0
Metalworking Machinery	42	Connecticut	6	450.0
Preserved Fruits and Vegetables	41	Georgia	8	1510.0
Household Audio and Video	38	New Jersey	11	111.1
Optical Instruments	37	Massachusetts	10	311.1
Electrical Distribution Equipment	35	California	10	125.0
Engineering Equipment	34	New Jersey	11	325.0
Motor Vehicles and Equipment	30	Michigan	21	2685.7

N indicates the total number of innovations for an industry. n indicates the number of innovations for a state.

State governments have traditionally had a central role in the provision of infrastructure, although this is typically interpreted as physical infrastructure. For innovative industries, scientific and technical infrastructure is as important as physical infrastructure. States are in a unique position to pull together the diverse public institutions and private entities that play a role in the innovation process. To develop an emerging expertise, states can strategically deploy funds to local areas and industries.

There is no single, successful formula for states to use in designing programs and initiatives that will promote innovation and technology transfer (Plosila 1987). Practical examples are provided by California's Silicon Valley, Massachusetts's Route 128 and North Carolina's Research Triangle Park. In many ways, these examples are unique: Silicon Valley relied on strong university-industry interaction; Route 128 benefitted from a strong entrepreneurial tradition; and, in large part, Research Triangle Park succeeded because of the direct efforts of the state government. Perhaps the common thread in these examples is that the initial event which sparked the development of the industry in the geographic areas was translated into a sustained advantage for an industry and a region. The results presented here suggest that states may succeed in fostering economic development if they understand and capitalize on their role in building and in augmenting the knowledge resources which enhance the commercialization process.

States are very aware that their policies must respond to the needs of innovative industries. Recently, state emphasis has visibly shifted from the provision of a favorable business climate to the provision of a favorable economic climate (John 1987). A favorable business climate was interpreted to mean a docile labor force, low taxes, and flexible regulations. By definition, a favorable economic climate encompasses a broader set of concerns such as the availability of skilled labor, the existence of an adequate infrastructure for economic growth, and the establishment of high quality public services. Unfortunately, these two concepts are at odds because education, infrastructure, and public services are expensive and conflict with the low tax objective associated with a favorable business climate. Moreover, the definition of high-quality infrastructure requires that attention be paid to the needs of industry located in the state, such as specific university research programs, technology transfer initiatives and shared scaling-up facilities.

An additional concern to address is the lengthy time frame necessary for state efforts to succeed. North Carolina's Research Triangle provides a case in point.[4] The Research Triangle Park is the direct result of state government intervention and provides insight into the time commitment required to increase regional innovative capacity (Luger 1984). The idea behind Research Triangle Park was to build on a commitment to higher-education in order to establish a base for economic development. The Research Triangle Park was established in the center of an imaginary triangle which links the University of North Carolina at Chapel Hill, Duke University in Durham, and North Carolina State in Raleigh. The goal of this joint-venture was to enhance the standing of the three universities by linking their resources. The preliminary planning began in 1952 and the growth of the park was slow.[5] It is only recently that Research Triangle Park is considered an economic development success (Malecki 1991). Many of the employees at the Research Park are graduates of the three universities. In this way, any state's funds that are invested in human capital are retained and productively employed in the state. Most importantly, the efforts to build Research Triangle transcend several different political administrations. A commitment to the project was maintained for approximately fifteen years before the project became a commercial success. Such patience is seldom seen in the political arena that routinely demand results in two or four-year election cycles.

To a large extent, the abstract concept of technological infrastructure relies on the state educational system. States provide a large part of the financing for education and set standards for curriculum and achievement. Consequently, state government and state legislatures exert a large measure of fiscal control over higher education, the source of university research, and skilled technical labor.[6] A state's scientific and technical infrastructure promotes innovative activity so that economic growth can occur. States with better-developed innovative inputs, such as California, Massachusetts and New Jersey, are able to draw upon the established institutions and resources which already exist. Other states, notably Texas, demonstrate that state initiatives can build and strengthen innovative inputs as a mechanism to foster innovative activity.

Innovations within the industries listed in Table 6-1 exhibit a high degree of geographic clustering. The magnitude of the location quotients suggests that the prominent states have already developed a comparative advantage for that industry. This table lists many of the well-known innovative states. Other states, such as Ohio,Wisconsin and Georgia, emerge on this list. The industries listed are already a viable concern in these states. More importantly, these industries are the source of new product innovations which provide an emerging expertise which may lead to new rounds of state economic development.

Augmenting the State Knowledge-bases

This section considers state policies with regard to each of the innovative inputs: universities research, industrial R&D, related-industry presence and specialized business services. The objective of this section is to provide an analysis of state policy that affects the development of the innovative inputs.[7]

Our results indicate a strong association between university research expenditures and the location of product innovation. Universities provide two important complementary resources for innovative activity, scientific knowledge and skilled workers (Nelson 1986). The premise of the American university is that research is better when it includes the education and training of students. Education and training are more rigorous and creative when they are done in the context of an active research program (Rosenzweig and Turlington 1982).

The empirical results suggest that university research expenditures at the departmental level reflect the presence of related industry presence and R&D activities. The expenditures for university research increase in departments that are linked to related industry's manufacturing and R&D activity. The decreased importance of state and industrial funding of university research in the post-war period may hinder the innovation process. In 1935, industry funding was the largest source of university research funding (See Table 6-2). Over the post-War period, federal funding of university research has grown rapidly. The Second World War demonstrated the military value of university research. The Federal response to the ensuing cold war was to fund university research as a means of deliberately strengthening basic scientific research, and in turn, U.S. military prowess. Post-war science also became more expensive. Research in the emerging areas of high-energy physics, biomedical research, and engineering sub-fields is beyond the budgetary capacity of most universities and is outside the reasonable sphere of interest of the private sector (Rosenzweig and Turlington 1982).

The increased Federal financing of university research in the post-war period has not been matched by a similar increase in support from either state governments or from industry. The fact that university research provides a public good widely-disseminated throughout the country provides a justification for federal funding. The realization that benefits of university research are captured within-state boundaries makes an efficiency argument for increased state funding.

Table 6-2: Support for Academic Research by Sector: Constant 1982 dollars			
Year	1935	1960	1987
Federal	$67 (16.0%)	$1,157 (63.0%)	$ 6,196(60.6%)
State	$58 (14.0%)	$ 273 (13.2%)	$ 862 (8.4%)
Industry	$277 (66.0%)	$ 129 (6.2%)	$ 657 (6.4%)
	$420	$2,077	$10,218

Source: Mowery and Rosenberg (1991) provided data for 1935. The more recent years are from National Science Board (1989). These numbers do not sum to 100 due to the exclusion of institutions' own funds and all other sources.

Historically, the role of state funds for the university research system was greater during the pre-1940 period than the post-war period. A sizeable portion of pre-war state funds were devoted to extension activities, including the testing and support for the dissemination of best-practice techniques to local conditions. The politics of state funding may have meant that the research of U.S. public universities was more closely aligned with commercial opportunities within the state. During the pre-war time period, the state university systems often introduced new programs in emerging sub-fields of engineering as soon as the requirements of the local economy became clear (Mowery and Rosenberg 1991). The increased reliance on federal funding may have decreased university responsiveness to individual state requirements by removing the research agenda from state control.[8]

There are several examples of state programs which focus on selected fields as a targeted approach to foster innovation (Malecki 1987). Arizona and North Carolina have established state-of-the-art micro-electronic research centers with a mixture of state and industrial support. New York, New Jersey, and Ohio have sponsored advanced technology centers devoted to one technology. These programs also recognize the importance of critical mass by concentrating their resources in one location. For example, Ohio has established centers which focus on local core-competency and expertise: manufacturing process technology at the University of Cincinnati; welding research and metallurgy at Ohio State University; and, polymers at the University of Akron.

There is evidence which raises questions about Jaffe's finding that a dispersed, public university system will decrease industry research funds (1989). The aggregate patterns upon which these results are based may be misleading. Industry funding is a more prominent source of funding at smaller, specialized regional universities. For example, the University of Akron received 35.8% of its 1987 research budget from industry. This is in contrast with a share of 5.5% for industries funding for the top 25 institutions, ranked in terms of total academic research funds received. In general, the lower the ranking of the institution, in terms of total research funds received, the greater the percentage of industry funding (National Science Board 1989). This pattern may reflect a tendency of smaller regional schools to focus on the needs of local industry and participate in more cooperative research with industry.

To a large extent, industrial R&D investment is determined by scientific opportunity and expectation about the potential product market. Universities provide a source of critical information that will attract firms to locate R&D facilities nearby for a high opportunity technology which is evolving rapidly. A university that cultivates an expertise relevant to a specific industry's needs may attract related industry R&D to the state in order to gain access to that expertise. Texas provides an example of one state's successful effort to build up the university system. Once accomplished, these efforts translated into an increased number of industrial R&D labs in the state.

There is no denying that university and industrial cooperative research relationships can occur at great distance. But, to suggest that this indicates that location does not matter ignores the nature of the process by which local areas develop expertise. Firms fund research at those universities which have an expertise in a technology in which the company sees a potential commercial pay-off. The increased funding allows researchers and graduate students to sharpen their expertise and develop a further competitive advantage in the new technology. Regardless of the location of the sponsoring company, contractual research creates spill-overs which will be realized in the area of the university in subsequent time periods. It is no surprise that the highest numbers of new biotechnology start-ups are realized by states that have high levels of university funding in related disciplines (Kenney 1986). As researchers work with a new technology, ideas for commercial application are likely to occur. These ideas will be shared with the long-distance sponsor, however more frequent contact may foster more fruitful information exchanges. The benefits of the increased expertise funded by long distance sponsors may be captured by the local area.

In the local arena, business services provide the closest link to the market. The presence of business services may be critical if an area is to realize the benefits of the other innovative inputs, however this aspect of the innovation process is frequently ignored. State programs aimed at technology transfer may be more successful if business services are provided. A 1988 survey for the National Association of State Development Agencies (NASDA) found that although 28 states offer some type of management-assistance program, these programs accounted for only 2.3% of total expenditures by states on technology programs (Atkinson, 1988). In general, programs such as technical assistance, technology-transfer and diffusion of technology programs which provide links to the market amounted to less than 10% of total state expenditures. The author of this survey, Robert Atkinson concludes that, "To date, it appears that states have been much more willing to support early stages of research rather than try to assist existing firms to become more innovative" (Atkinson 1988: 32).

Many states attempt to encourage and nurture new firms as an economic development policy. The empirical evidence suggests that small firms benefit more from the presence of specialized business services. The cost of specialized services is prohibitive for small, start-up firms. For this reason, the state of Maryland is attempting to raise public money for a Bio-testing facility which would allow small biotechnology firms to test, evaluate and scale-up their new products. Proponents argue that this type of center would be an important resource in facilitating the growth of biotechnology firms in the state. The cost of this type of facility is beyond the ability of any single private firm. The scientific and technical infrastructure is enhanced if state government provides such a facility.

The last element to be considered is the related industry presence, especially as it embodies tacit knowledge which accompanies a familiarity with a technology. Not only is industry presence strongly associated with commercially-viable product innovations but it also influences the location of industrial R&D and the location of university research. These results suggest the importance of industry presence as a base upon which economic development efforts can build. Kolderie and Blazar (1988) document Minnesota's efforts to deliberately create scale economies for promising industries.

An important, but often-overlooked, source of tacit knowledge is contained within the manufacturing work-force. In contrast to the once-traditional organization of American manufacturing that organized work into unskilled components, technological innovation requires a work-force that can learn new skills and find new solutions to problems that arise on the job. This gains greater recognition as corporations change their internal organization to recognize the importance of the inputs of production workers. The importance of skilled workers to economic development should not be under-estimated. In contrast to the mobility of financial capital, this type of human capital is less mobile and provides a valuable resource which permits industry to innovate. State education policy provides the means to guarantee a supply of these skilled workers.

In conclusion, this study demonstrates the ways that geography is important to innovation and to the commercialization process. The evidence presented here suggests that the innovation process is facilitated by location. Innovations exhibit a pronounced tendency to cluster in states which contain concentrations of the knowledge resources which provide inputs to the innovation process. The location of the innovative inputs is mutually-reinforcing and, in this sense, comprise an integrative infrastructure for innovation. These findings enhance our understanding of how areas develop a comparative advantage for innovation. This comparative advantage does not rest on natural resource endowments or the availability of low cost labor. It is provided by the location of specialized knowledge resources -- the underlying technological infrastructure -- which is required to organize and facilitate the innovation process.

....

Chapter Notes:

1. Michael Porter (1990) argues that nations develop an international competitive advantage for industries based on a specialized scientific base combined with strong localized competition. New product innovations are an important component of this competitive advantage. Although our results are empirical and sub-national, the basic conceptualization holds.

2. Dertouzos et al, (1988); and Florida and Kenney (1990) provide comprehensive reviews of evidence related to this point.

3. See Link and Bauer (1990); Johnson and Edwards (1987); and, Badaracco (1991).

4. Research Triangle Park is provided as an example of direct government involvement in planning an innovative region. See Luger (1984, 1985) and Whittington (1985) for more information on Research Triangle Park.

5. The Research Triangle Park was established in the center of the triangle which linked the three universities in 1959, after seven years of planning. The Monsanto Company located an industrial R&D operation in the Research Park and for the next six years there were no other leases signed. In 1965, IBM became a tenant at the Park, followed by Data General, DuPont, General Electric. By 1987, the Park had forty-seven occupants, employing 27,000 with an annual payroll over half a billion dollars.

6. For many years, midwestern governors have complained that their states subsidize the training of engineers and scientists who subsequently move to one coast or the other. This migration is prompted by the relatively poorer employment opportunities in the states which provide the training. A state economic development initiative which fosters innovative activity may help to retain this important resource.

7. No attempt is made to describe state programs. There are several excellent recent studies which describe and catalogue state technology policy. See Atkinson (1988), Fosler (1988), Osborne (1988).

8. An international comparison of fields of academic research found that the U.S. spends over 50% of academic research in the life sciences (Mowery and Rosenberg 1991). This is a higher percentage than any of our trading partners. In contrast, France emphasizes the physical sciences and Japan strongly supports academic research in engineering.

Appendix: Detailed Data Description and Tables

The purpose of this appendix is to provide more detailed information on the SBA Innovation Citation Data and to give full details for the abbreviated tables found in Chapter 3. This appendix also provides a listing of the geographic concentration variable which is used in the model estimation.

SBA Innovation Citation Data

The Innovation Citation data was collected under a contract by the U.S. Small Business Administration (SBA) with the Futures Group. The final report for this contract, and the basis for this description, is provided by Edwards and Gordon (1984). The data provides an enumeration of innovations that were introduced to the U.S. market in the year 1982. The goal of the contract was to construct a comprehensive database for the study of innovation.

The data was collected via an intensive review of the new product announcement sections of over 100 trade journals and publications. Additional data on the innovations and the innovating firms were collected by conducting telephone interviews with the editors and the companies. Published, secondary sources were used to provide and to verify information on the companies.

The innovation data include the following items:

a. the model name of the innovation: e.g. Polatrol 3258;
b. the product name of innovation: e.g. High Speed Computer;

c. description of the innovation;
d. year of introduction;
e. year of invention;
f. innovation type: either product, process, service or management process;
g. innovation significance: judgmental assessment about the innovation's impact;
h. market characteristics: market size and market aggregation;
i. source of funding;
j. origin of the technology which led to the innovation;
k. name and state location of innovating entity;
l. SIC code of innovating entity;
m. name and state location of innovating enterprise, if different from the innovating entity;
n. the number of employees in the innovating enterprise at the time of innovation;
o. annual sales of innovating enterprise;
p. date of incorporation of the innovative enterprise.

The database consists of 8,074 innovation citations. Of these innovation citations, 4,476 were product innovations in manufacturing industries. Of these innovations, 4200 have information on the location of the innovating establishment.

For our purposes, the state identifier of the establishment is used to investigate the spatial patterns of innovation. It is anticipated that some innovations are developed by subsidiaries or divisions of companies with headquarters in other states. Since headquarters may announce new product innovations, the SBA data discriminate between the location of the innovating establishment and the location of the innovating entity. The site responsible for the major development of the innovation is known as the establishment. The parent company or headquarters is known as the entity. For example, Intel Corporation introduced a 16-Bit Micro-Controller (Model Number 8096). The major development was done by a division of Intel in Arizona while Intel's headquarters are in California. In this case, the state of the establishment is Arizona and the state of the entity is California: the innovation would be attributed to the state of Arizona.

Detailed Tables

Table A-1: Distribution of Innovations by Industry Groups

Industry Group	Innovations
Food and Kindred Products	100 (2.6%)
Lumber and Furniture	75 (1.8%)
Chemicals (excluding drugs)	159 (3.8%)
Drugs[1]	133 (3.2%)
Rubber and Plastics	60 (1.4%)
Primary Metals	17 (0.4%)
Fabricated Metal Products	201 (4.8%)
Machinery (excluding office)	438 (10.4%)
Computers and Office Equipment[2]	954 (22.7%)
Industrial Electrical Equipment[3]	402 (9.6%)
Household Appliances[4]	62 (1.5%)
Communications Equipment[5]	376 (9.0%)
Motor Vehicles and Transportation Equipment[6]	48 (1.1%)
Aircraft and Engines	16 (0.4%)
Measuring and Controlling Devices[7]	668 (15.9%)
Medical Instruments[8]	228 (5.4%)
General Instruments	146 (3.5%)

---

[1] SIC 283.
[2] SIC 357.
[3] Includes SIC 361, 362, 364 and 367.
[4] Includes SIC 363, 365 and 369.
[5] SIC 366.
[6] Includes SIC 371, 373, 374, 375 and 379.
[7] SIC 382
[8] SIC 384.

Table A-2: Which States are the most innovative?		
State	Number	Innovations per 100,000 Manufacturing Workers
New Jersey	426	52.33
Massachusetts	360	51.87
California	974	46.94
New Hampshire	33	30.84
New York	456	29.48
Minnesota	110	28.65
Connecticut	132	28.51
Arizona	41	27.70
Colorado	42	22.46
Delaware	15	21.13
National	4200	20.34
Rhode Island	24	18.46
Pennsylvania	245	18.28
Illinois	231	18.16
Texas	169	16.14
Wisconsin	86	15.61
Washington	48	15.38
Ohio	188	15.00
Florida	66	14.60
Oregon	32	14.48
Vermont	6	12.24
Idaho	6	12.00
Utah	11	11.83
D.C.	2	11.76
Maryland	28	11.11
Michigan	112	11.02

State	Number	Innovations per 100,000 Manufacturing Workers
Oklahoma	20	10.15
Georgia	53	10.10
Nebraska	9	9.47
Virginia	38	9.09
New Mexico	3	9.09
Missouri	36	8.20
Iowa	20	8.16
Kansas	15	7.77
Indiana	49	7.49
Nevada	1	5.56
North Carolina	38	4.63
South Carolina	18	4.60
Tennessee	20	4.07
Hawaii	1	3.70
West Virginia	4	3.48
Maine	4	3.45
Kentucky	9	3.33
Arkansas	5	2.51
Louisiana	5	2.39
Mississippi	4	1.89
Alabama	5	1.43

Source: Numbers of manufacturing workers are from the *1982 Census of Manufacturers*.

Table A-3: How do states compare on various measures of innovative activity ?					
State	Number of Innovations	Innovations per 100,00 Workers	Number of Patents	Patents per 100,000 Workers	% High Tech Workers
Alabama	5	1.43	69	19.77	14
Arizona	41	27.70	245	70.20	41
Arkansas	5	2.51	39	11.17	20
California	974	46.94	3230	925.50	37
Colorado	42	22.46	332	95.13	35
Conn	132	28.51	1090	312.32	39
Delaware	15	21.13	N.A.	N.A.	17
D.C.	2	11.76	N.A.	N.A.	0
Florida	66	14.60	424	121.49	31
Georgia	53	10.10	192	55.01	10
Idaho	6	12.00	N.A.	N.A.	14
Illinois	231	18.16	2451	702.29	28
Indiana	49	7.49	784	224.64	21
Iowa	20	8.16	243	69.63	23
Kansas	15	7.77	145	41.55	37
Kentucky	9	3.33	209	59.89	17
Louisiana	5	2.39	163	46.70	29
Maine	4	3.45	N.A.	N.A.	9
Maryland	28	11.11	N.A.	N.A.	30
Mass.	360	51.87	1339	383.67	33
Michigan	112	11.02	1725	494.27	15
Minnesota	110	28.65	628	179.94	27
Miss.	4	1.89	N.A.	N.A.	11
Missouri	36	8.20	423	121.20	20
Nebraska	9	9.47	50	14.33	21
Nevada	1	5.56	N.A.	N.A.	23
New Hampshire	33	30.84	N.A.	N.A.	18

Table A-3: How do states compare on various measures of innovative activity ?					
State	Number of Innovations	Innovations per 100,00 Workers	Number of Patents	Patents per 100,000 Workers	% High Tech Workers
New Jersey	426	52.33	2631	753.87	30
New Mexico	3	9.09	N.A.	N.A.	26
New York	456	29.48	2857	818.62	22
N Carolina	38	4.63	327	93.70	12
Ohio	188	15.00	1999	572.78	22
Oklahoma	20	10.15	453	129.80	33
Oregon	32	14.48	N.A.	N.A.	12
Penn.	245	18.28	2329	667.34	23
Rhode Is.	24	18.46	94	26.93	16
S Carolina	18	4.60	N.A.	N.A.	19
Tennessee	20	4.07	N.A.	N.A.	17
Texas	169	16.14	N.A.	N.A.	32
Utah	11	11.83	103	29.51	31
Virginia	38	9.09	297	85.10	26
Washington	48	15.38	N.A.	N.A.	17
W Virginia	4	3.48	N.A.	N.A.	27
Wisconsin	86	15.61	634	181.66	24

Table A-4: Number of Innovations by State and Two-digit SIC Industry

State	Number of Innovations by Two-digit SIC code										
	n	20	24 & 25	28	30	33	34	35	36	37	38
National Total	4200	110	75	730	60	17	201	954	840	66	1042
Alabama	5		1					3			1
Arizona	41			1			2	7	21		9
Arkansas	5	1					1		4		
California	974	8	10	51	7	1	26	365	288	18	187
Colorado	42			6	1			16	9		9
Connecticut	132	3	1	29	3	1	5	29	24	3	33
Delaware	15			6	1				2		3
D.C.	2								1		
Florida	66	1	1	14			2	17	17		13
Georgia	53	12		5	2		3	9	8	1	11
Hawaii	1							1		1	
Idaho	6	1						2	1		2
Illinois	231	5	11	61	2		18	35	31	1	56
Indiana	49			16		1	2	4	11	2	9
Iowa	20	2	3	2			2	5	1	1	4
Kansas	15		1	2	1			5	3	1	1
Kentucky	9			4					2		1
Louisiana	5				1			1	1		2
Maine	4			2					1		
Maryland	28			5			2	10	5		6
Massachusetts	360	13	5	42	5	1	15			3	120
Michigan	112	1	8	26	3		7	14	9	21	

Table A-4: Number of Innovations by State and Two-digit SIC Industry

State	Number of Innovations by Two-digit SIC code										
	n	20	24 & 25	28	30	33	34	35	36	37	38
National Total	4200	110	75	730	60	17	201	954	840	66	1042
Minnesota	110	61	3	19	3		8	24	14		28
Mississippi	4								3		
Missouri	36	5	2	11			2	7	2		6
Nebraska	9			2				1	3		3
Nevada	1										1
New Hampshire	33		2	3			3	18	2	1	4
New Jersey	426	6		124	5	4	19	65	58		131
New Mexico	3						1		1	1	
New York	456	20	8	73	8	2	20			2	147
North Carolina	38		1	16			3	2	8	1	6
Ohio	188	3	6	51	9	3	21	16	27	3	44
Oklahoma	20	1		1	1		2	4	4		7
Oregon	32	1		4			1	10	7		8
Pennsylvania	245	14	5	71	2	1	16	34	30		66
Rhode Island	24			6		1	1	3	8		5
South Carolina	18		2	8					2	1	5
Tennessee	20			6	1		1		1	1	7
Texas	169	4	2	25	2	1	10	47	35	2	37
Utah	11							8	1		2
Vermont	6							5			
Virginia	38		1	9	3			6	7		10

Table A-4: Number of Innovations by State and Two-digit SIC Industry

State	Number of Innovations by Two-digit SIC code										
	n	20	24 & 25	28	30	33	34	35	36	37	38
National Total	4200	110	75	730	60	17	201	954	840	66	1042
Washington	48		1	5			1	15	18	1	7
West Virginia	4						1				2
Wisconsin	86	1	1	23			6	7	17	1	29

Note: SIC 20: Food and Kindred Products; SIC 24&25: Wood and Furniture Products; SIC 28: Chemicals and Allied Products; SIC 33: Fabricated Metals; SIC 34: Fabricated Metal Products; SIC 35: Industrial and Commercial Machinery and Computer Equipment; SIC 36: Electronic and Other Electrical Equipment; SIC 37: Transportation Equipment; SIC 38: Measuring, Analyzing and Controlling Instruments.

Table A-5: Distribution of Three-Digit Innovations by State

State		Count	% of Innovations	% of State	Location Quotient
Computers 357		954			
	California	365	38.3	37.6	164.91
	Massachusetts	82	8.6	22.8	100.00
	New York	77	8.1	17.0	74.56
	New Jersey	65	6.8	15.4	67.54
	Texas	47	4.9	27.8	121.88
Measuring/Controlling Instruments 382		668			
	California	134	20.1	13.8	86.25
	Massachusetts	94	14.1	26.1	163.13
	New York	63	9.4	13.9	86.88
	New Jersey	55	8.2	13.0	81.25
	Pennsylvania	54	8.1	22.0	137.50
	Illinois	39	5.8	17.0	106.25
Communication Equipment 366		376			
	California	116	30.9	11.9	132.22
	New York	45	12.0	9.9	110.00
	Massachusetts	34	9.0	9.4	104.44
	New Jersey	22	5.9	5.2	57.78
	Texas	19	5.1	11.2	124.44
Electrical Components 367		261			
	California	128	49.0	13.2	212.90
	Massachusetts	26	10.0	7.2	116.13
	New Jersey	15	5.7	3.6	58.06
	Texas	15	5.7	8.9	143.55
	New York	13	5.0	2.9	46.77

Table A-5: Distribution of Three-Digit Innovations by State

State		Count	% of Innovations	% of State	Location Quotient
Medical Instruments 384		228			
	New Jersey	57	25.0	13.5	245.45
	New York	51	22.4	11.2	203.64
	Ohio	13	5.7	7.0	127.27
	California	27	11.8	2.8	50.91
Drugs 283		133			
	New Jersey	52	39.1	12.3	410.00
	New York	18	13.5	4.0	133.33
	Pennsylvania	11	8.3	4.5	140.63
	Illinois	9	6.8	3.9	121.88
	Michigan	8	6.0	7.1	221.88

Table A-6: Geographic Concentration Index			
State	Index	State	Index
Alabama	0.20	Massachusetts	0.57
Arkansas	0.46	Michigan	0.57
Arizona	0.66	Minnesota	0.59
California	0.56	Mississippi	0.44
Colorado	0.66	North Carolina	0.21
Connecticut	0.21	Nebraska	0.38
Florida	0.23	New Jersey	0.83
Georgia	0.20	New York	0.49
Iowa	0.10	Ohio	0.33
Illinois	0.75	Oklahoma	0.30
Indiana	0.10	Pennsylvania	0.33
Kansas	0.27	Rhode Island	0.94
Kentucky	0.10	Utah	0.73
Louisiana	0.22	Virginia	0.18
		Wisconsin	0.37

Bibliography

Acs, Zoltan J., and David B. Audretsch. 1990. *Innovations and Small Firms*. Cambridge: The MIT Press.

Acs, Zoltan J., and David B. Audretsch. 1989. Patents as a Measure of Innovative Activity. *Kyklos* 42:171-180.

Acs, Zoltan J., and David B. Audretsch. 1988. Innovation in Large and Small Firms: An Empirical Analysis. *American Economic Review* 78:678-690.

Acs, Zoltan J., and David B. Audretsch. 1987. Innovation, Market Structure and Firm Size. *Review of Economics and Statistics* 69:567-575.

Acs, Zoltan J., David B. Audretsch, and Maryann P. Feldman. The Real Effects of Academic Research: A Comment. *American Economic Review* 82:363-367.

Acs, Zoltan J., David B. Audretsch, and Maryann P. Feldman. forthcoming. The Recipients of R&D Spill-overs: Firm Size and Innovation. *Review of Economics and Statistics.*

Acs, Zoltan J., David B. Audretsch, and Maryann P. Feldman. forthcoming. R&D Spill-overs and Innovative Activity. *Managerial and Decision Economics.*

Ady, R. M. 1986. Criteria Used For Facility Location Selection. In: *Financing Economic Development in the 1980s,* by N. Walzer and D.L. Chicoine. New York: Praeger.

Allen, Robert C. 1983. Collective Innovation. *Journal of Economic Behavior and Organization* 4:1-24.

Amemiya, Takeshi. 1984. Tobit Models: A Survey. *Journal of Econometrics* 24:3-61.

Amemiya, Takeshi. 1974. Multivariate Regression and Simultaneous Equation Models When the Dependent Variables are Truncated Normal. *Econometrica* 43:999-1012.

Amemiya, Takeshi. 1973. Regression Analysis When the Dependent Variable is Truncated Normal. *Econometrica* 42:997-1016.

Arora, Ashish, and Alfonso Gambardella. 1990. Complementarity and External Linkages: The Strategies of Large Firms in Biotechnology. *The Journal of Industrial Economics* 38:361-379.

Arrow, Kenneth. 1962. Economic Welfare and the Allocation of Resources for Invention. In *The Rate and Direction of Inventive Activity,* by National Bureau of Economic Research. Princeton: Princeton University Press.

Arrow, Kenneth. 1962b. The Economic Implications of Learning by Doing. *Review of Economic Studies* 29:155-173.

Arthur, W. Brian. 1990. Positive Feedbacks in the Economy *Scientific American.* 221:92-99.

Arthur, W. Brian. 1988. Competing Technologies: An Overview. In *Technical Change and Economic Theory,* by G. Dosi et al. London: Francis Pinter.

Arthur, W. Brian. 1986. Industry Location and the Importance of History. CEPR Paper no.43: Stanford University.

Arthur, W. Brian. 1990. Silicon Valley Locational Clusters: When Do Increasing Returns Imply Monopoly. *Mathematical Social Sciences* 19:235-251.

Arthur, W. Brian. 1988. Urban Systems and Historical Path Dependence. In *Cities and Their Vital Systems* by J. Ausubel and R. Herman, Eds. Washington, D.C.: National Academy Press.

Atkinson, Robert D. 1991. Some States Take the Lead: Explaining the Formation of State Technology Policies. *Economic Development Quarterly* 5:33-44.

Atkinson, Robert D. 1988. State Technology Development Programs *Economic Development Review.* 25:29-33.

Baily, M.N., and A.K. Chakrabarti. 1988. *Innovation and the Productivity Crisis*. Washington, D.C.: Brookings Institution.

Badaracco, Joseph L., Jr. 1991. *The Knowledge Link: How Firms Compete through Strategic Alliances*. Boston: Harvard Business School Press.

Barnet, Richard, and Ronald Muller. 1974. *Global Reach: The power of the multinational corporation*. New York: Simon and Schuster.

Beeson, Patricia, and Edward Montgomery. 1988. Universities and the Migration and Employment of High-tech Workers: The goose that laid the golden egg?" Working paper, Center for Regional Economic Issues: Case Western Reserve University.

Bergman, Edward M., and Harvey A. Goldstein. 1983. Dynamics and Structural Change in Metropolitan Economies. *Journal of the American Planning Association* 49:263-279.

Bernstein, Jeffrey I., and Ishaq M. Nadiri. 1988. Interindustry R&D Spill-overs, Rates of Return and Production in High-tech Industries. *The American Economic Review* 78:429-434.

Bluestone B., and B. Harrison. 1982. *The De-industrialization of America*. New York: Basic Books.

Boadway, Robin W., and David E. Wildasin. 1984. *Public Sector Economics, Second Edition,* Boston: Little, Brown and Company.

Bollinger, Lynn., Katherine Hope, and James M. Utterback. 1983. A Review of the Literature and Hypothesis on New Technology-Based Firms. *Research Policy* 12:1-14.

Bound, John, Clint Cummins, Zvi Griliches, Bronwyn H. Hall, and Adam Jaffe. 1984. Who Does R&D and Who Patents? In *R&D, Patents, and Productivity*, Zvi Griliches, Ed. Chicago: University of Chicago.

Brody, David, and Richard Florida. 1991. Falling Through the Cracks: The U.S. failure in active matrix display technology. Carnegie Mellon University: School of Urban and Public Affairs, Working Paper 91-10.

Brown, L.A. 1981. *Innovation Diffusion: A new perspective.* New York: Methuen.

Browning, J. 1980. *How to Select a Business Site.* New York: McGraw Hill.

Bush, Vannevar. 1945. *Science: The Endless Frontier.* Washington, D.C.: U.S.Government Printing Office.

Carlson, W. Bernard, and Michael E. Gorman. 1992. A Cognitive Framework to Understand Technological Creativity: Bell, Edison, and the telephone. In *Inventive Minds,* by Robert J. Weber and David N. Perkins, Eds. New York: Oxford University Press.

Carlson, Bo. 1992. The Rise of Small Business: Causes and Consequences. *Singular Europe: Economy and Polity of the European Community after 1992.* Ann Arbor: University of Michigan Press.

Carlson, Bo, and S. Jacobsson. 1991. What makes the Automation Industry Strategic. *Economics of Innovation and New Technology* 1:93-118.

Castells, M. 1985. High-technology, Economic Restructuring and the Urban-Regional Process in the United States. In *High-technology, Space and Society,* by M. Castells. Beverly Hills: Sage.

Chapman, Robert E., Marianne K. Clarke, and Eric Dobson. 1990. *Technology-Based Economic Development: A study of state and technical extension services.* Washington, D.C.: U.S. Government Printing Office.

Clark, Gordon. 1981. The Employment Relation and the Spatial Division of Labor: A hypothesis. *Annals of the Association of American Geographers* 71:412-424.

Clark, Gordon, Meric Gertler and John Whitman. 1986. *Regional Dynamics.* Boston: Allen and Unwin.

Clark, Marianne. 1986. *Revitalizing State Economies: A review of state economic development policies and programs.* Washington, D.C.: National Governors' Association.

Clark, N.G. 1972. Science, Technology and Regional Economic Development. *Research Policy* 1:296-319.

Coase, R.H. 1937. The Nature of the Firm. *Econometrica* 4:386-405.

Cobb, J.C. 1982. *The Selling of the South: The Southern crusade for industrial development 1936-1980*. Baton Rouge: Louisiana State University Press.

Coffey, William, J., and Michael Polese. 1987. Trade and Location of Producer Services: A Canadian perspective. *Environment and Planning A* 19:597-611.

Cohen, Wesley M., and Steven Klepper. 1990. Reprise of Size and R&D. mimeograph, Carnegie Mellon University.

Cohen, Wesley M., and Richard Levin. 1989. Empirical Studies of Innovation and Market Structure. In *Handbook of Industrial Organization, Volume II*, by R. Schmalensee and R.D. Willig. North Holland: Elsevier Science Pub. Co.

Cohen, Wesley M., and Richard C. Levin, and David Mowery. 1987. Firm Size and R&D Intensity: A re-examination. In *The Empirical Renaissance in Industrial Economics* by T.F. Bresnahan and R. Schmalensee. New York: Basil Blackwell.

Cohen, Wesley M., and Daniel A. Levinthal. 1990. Absorptive Capacity: A new perspective on learning and innovation. *Administrative Science Quarterly* 35:128-152.

Czmanski, Stanley and L. Ablas. 1979. Identification of Industrial Clusters and Complexes, *Urban Studies* 16:61-80.

Daniels, P. W. 1985. *Service Industries: A geographical appraisal.* New York: Methuen.

Dasgupta, Partha and Joseph Stiglitz. 1980. Industrial Structure and the Nature of Innovative Activity. *Economic Journal* 90: 266-293.

Davelaar, Evert, and Peter Nijkamp. 1989. The Role of the Metropolitan Milieu as an Incubator Center for Technological Innovations: A Dutch case study. *Urban Studies* 26:516-529.

David, Paul A. and Joshua L. Rosenbloom. 1990. Marshallian Factor Market Externalities and the Dynamics of Industrial Location *Journal of Urban Economics* 28:349-370.

David, Paul. 1985. Cliometrics and the QWERTY. *American Economic Review* 75:332-337.

DeBresson, C. 1989. Breeding Innovation Clusters: A source of dynamic development. *World Development* 17:1-16.

DeBresson, C., and F. Amesse. 1991. Networks of Innovators: A review and introduction to the issues. *Research Policy* 20:363-380.

Dertouzos, Michael L., R.K. Lester, and R.M. Solow. 1989. *Made In American: Regaining the productive edge*. Cambridge: The MIT Press.

Doheny-Farina, Stephen. 1992. *Rhetoric, Innovation, Technology: Case studies of technical communication in technology transfers*. Cambridge: The MIT Press.

Dorfman, Nancy S. 1987. *Innovation And Market Structure: Lessons from the computer and semiconductor industries*. Cambridge: Ballinger Publishing Company.

Dorfman, Nancy S. 1983. Route 128: The Development of a Regional High-technology Economy. *Research Policy* 12:299-316.

Dosi, Giovanni. 1988a. The Nature of the Innovative Process. In *Technical Change and Economic Theory* by G. Dosi, R. Nelson, G. Silverberg, C. Freeman and L. Soete, Eds. London:Pinter.

Dosi, Giovanni. 1988b. Sources, Procedures and Microeconomic Effects of Innovation. *Journal of Economic Literature* 36:1120-1171.

Edwards, Keith L., and Theodore J. Gordon. 1984. Characterization of Innovations Introduced on the U.S. Market in 1982. The Futures Group, U.S. Small Business Administration, Contract No. SBA-6050-0A-82.

Eisinger, Peter K. 1988. *The Rise of The Entrepreneurial State*. Madison: University of Wisconsin Press.

Ergas, H. 1987. Does High-technology Policy Matter. In *Technology and Global Industry* by H. Brooks and B. Guile. Washington, D.C.: National Academy Press.

Feldman, Maryann P. 1992. The Geography of Innovation: A cross sectional analysis of state level data. Unpublished doctoral dissertation. Carnegie Mellon University.

Feldman, Maryann P. forthcoming. An Examination of the Geography of Innovation. *Corporate and Industrial Change.*

Feldman, Maryann P. forthcoming. Knowledge complementarity and Innovation. *Small Business Economics.*

Feldman, Maryann P. forthcoming. The University and High-technology Start-Ups: The Case of Baltimore and Johns Hopkins University. *Economic Development Quarterly.*

Feldman, Maryann P., and Richard Florida. forthcoming. The Geography Sources of Innovation: Technological Infrastructure and Product Innovation in the United States. *Annals of the Association of American Geographers.*

Feller, Irwin. 1988. Evaluating State Advanced Technology Programs. *Evaluations Review* 12:232-252.

Feller, Irwin. 1984. Political and Administrative Aspects of State High-technology Programs. *Policy Studies Review* 3:460-466.

Ferguson, Charles H. 1988. From the People Who Brought You Voodoo Economics. *Harvard Business Review* 66:55-62.

Florida, Richard, and Martin Kenney. 1990. *The Breakthrough Illusion: Corporate America's Failure to Move from Innovation to Mass Production.* New York: Basic Books.

Florida, Richard, and Martin Kenney. 1988. Venture Capital, High-technology, and Regional Development. *Regional Studies* 22:33-48.

Florida, Richard, and Martin Kenney. 1988. Venture Capital and Technological Innovation in the U.S. *Research Policy* 17:119-137.

Florida, Richard, and Donald Smith. 1990. Venture Capital Formation, Investment and Regional Industrialization. *Annals of the Association of American Geographers* 83: 434-451.

Fosler, F.S. 1988. *The New Economic Role of American States.* New York: Oxford University Press.

Freeman, Christopher. 1991. Networks of Innovators: A Synthesis of Research Issues *Research Policy*. 20:499-514.

Freeman, Christopher. 1989. *The Economics of Industrial Innovation.* Cambridge: The MIT Press.

Freeman, Christopher, John Clark, and Luc Soete. 1982. *Unemployment and Technical Innovation.* Westport: Greenwood Press.

Friar, John, and Mel Horwitch. 1985. The Emergence of Technology Strategy: A new dimension of strategic management. *Technology in Society* 7:143-178.

Galbraith, John K. 1952. *American Capitalism: The Concept of Countervailing Power.* Boston: Houghton Mifflin Co.

Gilder, George. 1988. *Microcosm.* New York: Simon and Schuster.

Glasmeier, Amy K. 1990. High-Tech Policy, High-Tech Realities: The spatial distribution of high-technology industries in America. In *Growth Policy in the Age of High-technology: The role of regions and states* by Jurgen Schmandt and Robert Wilson. Boston: Unwin Hyman.

Glasmeier, Amy K. 1988. Factors Governing the Development of High-tech Industry Agglomerations: A tale of three cities. *Regional Studies* 22:287-301.

Glasmeier, Amy K. 1986. High-Tech Industries and the Regional Division of Labor. *Industrial Relations* 25:197-211.

Glasmeier, Amy K. 1985. Innovative Manufacturing Industries: Spatial incidence in the United States. In *High-technology, Space and Society* by M. Castells. Beverly Hills: Sage.

Glasmeier, Amy K., Peter Hall, and Ann Markusen. 1983. Recent Evidence in High-technology Industries' Spatial Tendencies: A preliminary investigation. *Technology, Innovation and Regional Economic Development.* Washington, D.C.: U.S. Government Printing Office.

Goddard, J.B. 1974. The Location of Non-manufacturing Ativities with Manufacturing Industries. In *Contemporary Industrialization*, by F.E. Hamilton. London: Longman.

Grabowski, Henry G., and Dennis C. Mueller. 1978. Industrial Research and Development, Intangible Capital Stocks and Firms Profit Rates. *Bell Journal of Economics* 9:328-343.

Greene, William H. 1989. *Econometric Analysis*. New York: Macmillian.

Grefsheim, Suzanne, Jon Franklin, and Diana Cunningham. 1991. Biotechnology Awareness Study, part 1: Where scientists get their information. *Bulletin of the Medical Library Association* 79:36-44.

Griliches, Zvi. 1990. Patent Statistics as Economic Indicator: A survey. *Journal of Economic Literature* 28:1661-1707.

Griliches, Zvi. 1986. Productivity, R&D, and Basic Research at the Firm Level in the 1970's. *American Economic Review* 76:141-154.

Griliches, Zvi. 1979. Issues in Assessing the Contribution of R&D to Productivity Growth. *Bell Journal of Economics* 10:92-116.

Griliches, Zvi. 1992. The Search for R&D Spill-overs. National Bureau of Economic Research Working Paper 3768.

Grossman, Gene M., and Elhanan Helpman. 1992. *Innovation and growth in the Global Economy*. Cambridge: The MIT Press.

Hagerstrand, T. 1967. *Innovation Diffusion as a Spatial Process*. Chicago: University of Chicago Press.

Hagerstrand, T. 1952. *The Propagation of Innovation*. Lund: Gleerup Lund Studies in Geography.

Hall, Bronwyn H. 1984. Software for the Computation of Tobit Model Estimates. *Journal of Econometrics* 24:215-222.

Hall, Bronwyn H., Zvi Griliches, and Jerry A. Hausman. 1986. Patents and R&D: Is there a lag? *International Economic Review* 27:265-302.

Hall, Peter, and Ann Markusen. 1985. *Silicon Landscapes*. Boston: Allen and Unwin.

Hall, Peter. 1985. Technology, Space and Society in Contemporary Britain. In *High-technology, Space and Society*, by M. Castells. Beverly Hills: Sage.

Harrison, Bennett. 1984. Regional Restructuring and 'Good Business Climate': The economic transformation of New England since World War II. In *Sunbelt/Snowbelt: Urban development and regional restructuring*, by L. Sawers and W.K. Tabb. New York: Oxford University Press.

Harrison, Bennett and Sandra Kanter. 1978. State Job Creation Business Incentives. *Journal of the American Planning Association* 44:424-435.

Howells, Jeremy. 1990. The Location and Organization of Research and Development: New horizons. *Research Policy* 19:133-146.

Howells, Jeremy. 1987. Developments in the Location, Technology and Industrial Organization of Computer Services: Some trends and research issues. *Regional Studies* 21:493-503.

Hughes, Thomas P. 1989. *American Genesis: A Century of Invention And Technological Enthusiasm 1870-1970*. New York: Penguin Books.

Hymer, Stephen H. 1979. The Multinational Corporation and the International Division of Labor. In *The Multinational Corporation: A radical approach*, by S. Hymer and R. Cohen. Cambridge: University Press.

Isard, W. 1956. *Location and Space-economy*. Cambridge: The MIT Press.

Jaffe, Adam B. 1989. Real Effects of Academic Research. *American Economic Review* 79:957-970.

Jaffe, Adam B. 1986. Technological Opportunity and Spill-overs of R&D. *American Economic Review* 76:984-1001.

Jaffe, Adam B., Manuel Trajtenberg, and Rebecca Henderson. 1993. Geographic Localization Of Knowledge Spill-overs As Evidenced By Patent Citations. *Quarterly Journal of Economics* 108:577-598.

John, DeWitt. 1987. *Shifting Responsibilities: Federalism in Economic Development.* Washington, D.C.: National Governor's Association.

Johnston, J. 1972. *Econometric Methods, Second Edition.* New York: McGraw-Hill.

Johnston, Robert F., and Christopher G. Edwards. 1987. *Entrepreneurial Science: New links between corporations, universities and government.* New York: Quorum Books.

Jorde, T.M., and David J. Teece. 1990. Innovation and Cooperation: Implications for competition and antitrust. *The Journal of Economic Perspectives* 4:75-96.

Judge, G.G. 1985. *The Theory and Practice of Econometrics.* New York: John Wiley and Sons.

Justman, Moshe. Spatial Dimensions of Industrial Linkages. mimeograph, Center for Regional Economic Issues, Case Western Reserve University.

Kamien, Morton I., and Nancy L. Schwartz. 1975. Market Structure and Innovation: A survey. *Journal of Economic Literature* 13:1-37.

Kennedy, Peter. 1987. *A Guide to Econometrics, Second Edition.* Cambridge: The MIT Press.

Kenney, Martin. 1986. *Biotechnology: The university industry complex.* New Haven: Yale University Press.

Kline, Stephen J., and Nathan Rosenberg. 1987. An Overview of Innovation. In *The Positive Sum Strategy,* by R. Landau and N. Rosenberg. Washington, D.C.: National Academy Press.

Krugman, Paul. 1991a. *Geography and Trade.* Cambridge: The MIT Press.

Krugman, Paul. 1991b. Increasing Returns and Economic Geography. *Journal of Political Economy* 99:483-499.

Kutay, Aydan. 1988. Technological Change and Spatial Transformation in an Information Economy. *Environment and Planning A* 20:707-718.

Lambright, W. Henry, and Albert Teich. 1989. Science, Technology and State Economic Development. *Policy Studies Journal* 18:135-147.

Landau, Ralph, and Nathan Rosenberg. 1987. *The Positive Sum Strategy*. Washington, D.C.: National Academy Press.

Landes, David S. 1969. *The Unbound Prometheus*. Cambridge: Cambridge University Press.

Langbein, Laura Irwin, and Allan J. Lichtman. 1978. *Ecological Inference*. Beverly Hills: Sage University Press.

Lazonick, William. 1991. *Business organization and the Myth of the Market Economy*. New York: Cambridge University Press.

Lee, L. 1981. Simultaneous Equations Models with Discrete and Censored Dependent Variables. *Econometrica* 47:977-996.

Levin, Richard C. 1988. Appropriability, R&D Spending and Technological Performance. *American Economic Review* 78:424-428.

Levin, Richard C., Wesley M. Cohen, and David C. Mowery. 1985. R&D Appropriability, Opportunity, and Market Structure: New evidence on some Schumpeterian hypotheses. *American Economic Review* 75:20-14.

Levin, Richard C., Alvin K. Klevorick, Richard R. Nelson, and Sidney G. Winter. 1987. Appropriating the Returns from Industrial Research and Development. *Brookings Papers on Economic Activity* :783-831.

Levin, Richard C., and Peter C. Reiss. 1988. Cost-reducing and Demand-creating R&D with Spill-overs. *Rand Journal of Economics* 19:538-556.

Link, Albert N., and Laura L. Bauer. 1989. *Cooperative Research in U.S. Manufacturing: Assessing policy initiatives and corporate strategies*. Lexington: Heath-Lexington.

Link, Albert N., and John Rees. 1990. Firm Size, University Based Research and the Returns to R&D. *Small Business Economics* 2: 25-32.

Lucas, Robert E. 1988. On the Mechanics of Economic Development. *Journal of Monetary Economics* 22:3-42.

Lucas, Robert E. 1993. Making a Miracle. *Econometrica* 61:251-272.

Luger, M.I. 1984. Does North Carolina's high-technology Development Policy Work? *Journal of the American Planning Association* 50.

Luger, M.I. 1985. The States and High-technology Development: the case of North Carolina. In *High hopes for high-tech: microelectronics policy in North Carolina,* by D. Whittington, Ed. Chapel Hill: North Carolina University Press.

Lund, Leonard. 1986. *Locating Corporate R&D Facilities*. New York: The Conference Board.

Lundvall, Bengt-Ake. 1988. Innovation as an Interactive Process: User producer relations. In *Technical Change and Economic Theory,* by G. Dosi, Ed. London: Francis Pinter.

Machlup, Fritz. 1962. The Supply of Inventors and Inventions. In *The Rate and Direction of Inventive Activity,* by Richard Nelson, Ed. Princeton: Princeton University Press.

MacPherson, Alan. 1991. Interfirm Information Linkages in an Economically Disadvantaged Regions: An empirical perspective from metropolitan Buffalo. *Environment and Planning A,* 591-606.

MacPherson, Alan. 1988. New Product Development among Small Toronto Manufacturers: Empirical evidence on the role of technical service linkages. *Economic Geography* 64:61-75.

Maddala, G.S. 1983. *Limited Dependent Variables and Qualitative Variables in Econometrics.* Cambridge: Cambridge University Press.

Maddala, G.S. 1977. *Econometrics.* New York: McGraw-Hill.

McDonald, John F., and Robert A. Moffitt. 1980. The Uses of Tobit Analysis. *The Review of Economics and Statistics* 62:318-321.

McDowell, Robert W. 1981. Research Triangle Park: Its conception, birth, growth and economic significance. *North Carolina* 50:14-18.

Mahar, James, F., and Dean C. Coddington. 1965. The Scientific Complex - Proceed with Caution. *Harvard Business Review* 43:132-143.

Malecki, Edward J. 1990. Technological Innovation and Paths to Regional Economic Growth. In *Growth Policy in the Age of High-technology: The Role of Regions and States*, by Jurgen Schmandt and Robert Wilson, Ed. Boston: Unwin Hyman.

Malecki, Edward J. 1987. Hope or Hyperbole: High-tech and Economic Development. *Technology Review* 90:45-51.

Malecki, Edward J. 1987. The R&D Location of the Firm and Creative Regions: A survey. *Technovation* 6:205-222.

Malecki, Edward J. 1986. Research and Development and the Geography of High-Technology Complexes. In *Technology, Regions and Policy*, by John Rees, Ed. Totowa: Rowman and Littlefield.

Malecki, Edward J. 1985a. Industrial Location and Corporate Organization in High-technology Industries. *Economic Geography* 61:345-369.

Malecki, Edward J. 1985b. Public Sector Research and Development and Regional Economic Performance in the United States. In *The Regional Economic Impact of Technological Change*, by A.T. Thwaites and R.P. Oakley, Ed. London: Pinter.

Malecki, Edward J. 1981. Government Funded R&D: Some regional economic implications. *Professional Geographer* 33:72-82.

Malecki, Edward J. 1981. Product Cycles, Innovation Cycles and Regional Economics Change. *Technological Forecasting and Social Change* 19:291-306.

Malecki, Edward J. 1980. Dimensions of R&D Location in the United States. *Research Policy* 9:2-22.

Malecki, Edward J. 1980. Firm Size, Location and Industrial R&D: A disaggregated analysis. *Review of Business and Economic Research* 16:29-42.

Malecki, Edward J. 1980. Technology and Regional Development: A survey. *International Regional Science Review* 8:89-125.

Malecki, Edward. 1990. *Technology and Economic Development.* Essex: Longman Scientific and Technical.

Malecki, Edward. 1981. Science, Technology, and Regional Economic Development: Review and Prospects. *Research Policy* 10:312-314.

Malecki, Edward J., and S.L. Bradbury. 1992. R&D Facilities and Professional Labour: Labour Force Dynamics in High Technology. *Regional Studies* 26:123-136.

Mansfield, Edwin J. 1991. Academic Research and Industrial Innovation. *Research Policy* 20:1-12.

Mansfield, Edwin J. 1988. Industrial R&D in Japan and the United States: A comparative study. *American Economic Review* 78:223-228.

Mansfield, Edwin J. 1984. Comment on Using Linked Patent and R&D Data to Measure Interindustry Technology Flows. In *R&D, Patents, and Productivity*, by Zvi Griliches Ed. Chicago: University of Chicago.

Markusen, Ann. 1985. *Product Cycles, Oligopoly and Regional Development.* Cambridge: The MIT Press.

Markusen, Ann, Peter Hall, and Amy Glasmeier. 1986. *High-tech America: The what, how, where and why of sunrise industries.* Boston: Allen and Unwin.

Markusen, Ann, and Karen McCurdy. 1989. Chicago's Defense Based High-technology: A case study of the 'Seedbeds of Innovation' hypothesis. *Economic Development Quarterly* 3:15-31.

Marshall, A. 1890. *Principles of Economics.* London: Macmillian.

Marshall, A. 1949. *Elements of Economics of Industry.* London: Macmillan.

Marshall, J.N. 1982. Linkages between Manufacturing Industry and Business Services. *Environment and Planning A* 14:1523-1540.

Marshall, J.N., P. Damesick, and P. Wood. 1987. Understanding the Location and Role of Producer Services in the United Kingdom. *Environment and Planning A* 19:575-595.

Miller, Roger, and Marcel Cote. 1985. Growing the Next Silicon Valley. *Harvard Business Review* 63:114-123.

Mowery, David. 1983. Industrial Research, Firm Size, Growth and Survival, 1921-1946. *Journal of Economic History* 43:953-980.

Mowery, David, and Nathan Rosenberg. 1991. The U.S. National Innovation System. Mimeograph.

Mowery, David, and Nathan Rosenberg. 1989. *Technology and the Pursuit of Economic Growth.* New York: Cambridge University Press.

Mowery, David, and Nathan Rosenberg. 1979. The Influence of Market Demand Upon Innovation: A critical survey of several recent empirical studies. *Research Policy* 8:102-153.

National Science Board. 1989. *Science and Engineering Indicators.* Washington, D.C.: U.S. Government Printing Office.

National Science Foundation. 1988. *Geographic Distribution of Industrial R&D Expenditures.* Washington, D.C.: U.S. Government Printing Office.

Nelson, Richard R. 1990. Capitalism as an Engine of Progress. *Research Policy* 19:193-214.

Nelson, Richard. 1986. Institutions Supporting Technical Advance in Industry. *American Economic Review* 76:186-189.

Nelson, Richard. 1982. The Role of Knowledge in R&D Efficiency. *Quarterly Journal of Economics* 97:453-470.

Nelson, Richard. 1988. Institutions Supporting Technical Change in the United States. In *Technical Change and Economic Theory*, by G. Dosi et al. Eds.

Nelson, Richard, and Sidney Winter. 1982. *An Evolutionary Theory of Economic Change.* Cambridge: Harvard University Press.

Nelson, Richard, and Sidney Winter. 1977. In Search of A Useful Theory of Innovation. *Research Policy* 6:36-76.

Nonaka, Ikujiro. 1991. The Knowledge-Creating Community. *Harvard Business Review* 69:96-104.

Norton, R., and J. Rees. 1979. The Product Cycle and the Spatial Decentralization of American Manufacturing. *Regional Science* 13:141-151.

Noyce, Robert. 1982. Competition and Cooperation: A prescription for the 80's. *Research Management* 25:13-17.

Noyelle, T.J., and T.M. Stanback. 1984. *The Economic Transformation of American Cities*. Totowa: Rowman and Allanheld.

Oakey, Ray. 1985. High-technology Industries and Agglomeration Economies. In *Silicon Landscapes*, by Peter Hall and Ann Markusen. Boston: Allen and Unwin.

Oakey, Ray. 1984. Innovation and Regional Growth in Small High-technology Firms: Evidence from Britain and the USA. *Regional Studies* 18:237-251.

Oakey, Ray. 1982. *High-technology Industry and Industrial Location*. Aldershot: Gower.

Oakey, Ray, and S.Y. Cooper. 1989. High-technology Industries, Agglomeration and the Potential for Peripherally Sited Firms. *Regional Studies* 23:347-360.

Osborne, David. 1988. *Laboratories of Democracy*. Boston: Harvard Business School Press.

Pavitt, Keith. 1984. Sectoral Patterns of Technical Change: Towards a taxonomy and a theory. *Research Policy* 13:343-373.

Pavitt, Keith, M. Robson, and J. Townsend. 1987. The Size Distribution of Innovating Firms in the U.K.: 1945-1983. *The Journal of Industrial Economics* 55:291-316.

Phillips, Bruce D. 1991. The Increasing Role of Small Firms in the High-Technology Sector: Evidence from the 1980s. *Business Economics* 26:40-47.

Piore, Michael J., and Charles F. Sabel. 1984. *The Second Industrial Divide*. New York: Basic Books.

Plosila, Walter H. 1987. State Technical Development Programs. *Forum for Applied Research and Public Policy* 30-38.

Pool, Robert. 1991. The Social Return of Academic Research. *Nature* 20:661.

Porter, M. 1990. *The Competitive Advantage of Nations*. New York: Free Press.

Powell, Walter W. 1987. Hybrid Organizational Arrangements: New form or transitional development? *California Management Review* 29:67-87.

Powell, Walter W. 1990. Neither Market nor Hierarchy: Network forms of organization. *Research in Organizations* 12:293-336.

Premus, R. 1982. *Location of High-technology Firms and Regional Economic Development*. Washington, D.C.: U.S. Government Printing Office.

Robertson, A. 1974. Innovation Management. *Management Decisions* 12:329-373.

Rogers, Everett M., and Judith K. Larsen. 1984. *Silicon Valley Fever: Growth of High-technology Culture*. New York: Basic Books.

Romer, Paul M. 1986. Increasing Returns and Long Run Growth. *Journal of Political Economy* 94:1002-1037.

Rosegrant, Susan, and David R. Lampe. 1992. *Route 128: Lessons from Boston's high-tech community*. New York: Basic Books.

Rosenberg, Nathan. 1982. *Inside the Black Box: Technology and economics*. New York: Cambridge University Press.

Rosenzweig, Robert M., and Barbara Turlington. 1982. *The Research Universities and Their Patrons*. Berkeley: University of California Press.

Rothwell, R., and W. Zegveld. 1982. *Innovation and Small and Medium Sized Firm*. London: Frances Pinter.

Saxenian, AnnaLee. 1991. The Origins and Dynamics of Production Networks in Silicon Valley. *Research Policy* 20:423-437.

Saxenian, AnnaLee. 1985. Silicon Valley and Route 128: Regional prototypes or historical exceptions? In *High-technology, Space and Society* by M. Castells. Beverly Hills: Sage.

Sayer, A., and R. Walker, R. 1993. *The New Social Economy*. Oxford: Basil Blackwell.

Scherer, F.M. 1991. Schumpeter and Plausible Capitalism. Harvard University, mimeo.

Scherer, F. M. 1983. The Propensity to Patent. *International Journal of Industrial Organization* 1:107-28.

Scherer, F.M. 1980. *Industrial Market Structure and Economic Performance*. Chicago: Rand McNally College Publishing.

Scherer, F.M. 1967. Market Structure and the Employment of Scientists and Engineers. *American Economic Review* 57:524-530.

Schmandt, Jurgen, and Robert Wilson. 1990. *Growth Policy in the Age of High-technology: The role of regions and states*. Boston: Unwin Hyman.

Schmenner, Roger W. 1982. *Making Business Location Decisions*. Englewood Cliffs: Prentice-Hall.

Schmookler, J. 1966. *Invention and Economic Growth*. Cambridge: Harvard University Press.

Schumpeter, Joseph. 1934. *The Theory of Economic Development*. Cambridge: Harvard University Press.

Schumpeter, Joseph. 1942. *Capitalism, Socialism and Democracy*. New York: Harper and Row.

Science Policy Research Unit. 1972. *Success and Failure in Industrial Innovation*. London: Center for the Study of Industrial Innovation.

Scott, Allen J. 1990. *Metropolis: From the division of labor to urban form*. Los Angeles: University of California Press.

Scott, Allen J. 1988. *New Industrial Spaces*. London: Pion.

Scott, Allen J., and David Angel. 1988. Global Assembly-operations of U.S. Semiconductor Firms. *Environment and Planning A* 20:1047-1067.

Scott, Allen J., and Michael Storper. 1988. High-technology Industry and Regional Development: A Theoretical Critique and Reconstruction. *International Social Science Journal* 112:215-232.

Scott, Allen, and Michael Storper. 1988. Work Organization and Local Labor Markets in an Era of Flexible Production. *International Social Science Journal* 129:573-591.

Shepherd, William G. 1979. *The Economics of Industrial Organization*. Englewood Cliffs: Prentice Hall.

Shimshoni, Daniel. 1966. Spin-Off Enterprises from a Large Government-Sponsored Laboratory. Unpublished doctoral dissertation, Sloan School, MIT, Cambridge.

Skeath, Susan J. 1988. Technological Innovation and Diffusion, Knowledge and the Potential for Regional Economic Development: An integrative survey. Working paper, Center for Regional Economic Issues, Case Western Reserve University.

Smilor, R., G. Kozmetsky, and D. Gibson. 1988. *Creating the Technopolis: Linking technology, commercialization and economic development*. New York: Ballinger Publishing Co.

Spence, A. M. 1984. Cost Reductions, Competition, and Industry Performance. *Econometrica* 52:101-121.

Stephan, Paula E., and Sharon G. Levin. 1992. *Striking the Mother Lode: The importance of age, place, and time*. New York: Oxford University Press.

Stiglitz, Joseph. 1987. Learning to Learn, Localized Learning and Technological Progress. In *Economic Policy and Technological Performance*, by P. Dasgupta and P. Stoneman. Cambridge: Cambridge University Press.

Struyk, R.J., and F.J. James. 1975. *Intra-Metropolitan Industrial Location*. Lexington: Heath.

Stohr, Walter. 1986. Regional Innovation Complexes. *Papers of the Regional Science Association* 59:29-44.

Storper, Michael, and Richard Walker. 1989. *The Capitalist Imperative: Territory, technology and industrial growth.* Oxford: Basil Blackwell.

Storper, Michael, and Richard Walker. 1984. The Spatial Division of Labor: Labor and the location of industries. In *Sunbelt/Snowbelt: Urban development and regional restructuring*, by Larry Sawers and William Tabb, Eds. New York: Oxford University Press.

Sveikauskas, Leo, John Gowdy, and Michael Funk. 1988. Urban Productivity: City size or industry size. *Journal of Regional Science* 28:185-202.

Sweeney, G.P. 1987. *Innovation, Entrepreneurs and Regional Development.* New York: St. Martins Press.

Tassey, G. 1991. The Functions of Technology Infrastructure in a Competitive Economy. *Research Policy* 20:329-343.

Teece, David, J. 1986. Profiting from Technological Innovation: Implications for integration, collaboration, licensing and public policy. *Research Policy* 15:285-305.

Teece, David J. 1980. Economics of Scope and the Scope of the Organization. *Journal of Economic Behavior and Organization* 1:223-247.

Thomas, M. 1985. Regional Economic Development and the Role of Innovation and Technological Change. In *The Regional Economic Impact of Technological Change*, by A.T. Thwaites and R.P. Oakley. New York: St. Martin's Press.

Thompson, Wilbur. 1968. *Preface to Urban Economics.* Baltimore: Johns Hopkins Press.

Thompson, Wilbur R. 1962. Locational Differences in Inventive Effort and Their Determinants. In *The Rate and Direction of Inventive Activity*, by Richard Nelson, Ed. Princeton: Princeton University Press.

Tobin, J. 1958. Estimation of Relationships For Limited Dependent Variables. *Econometrica* 26:24-36.

Trajtenberg, Manuel. 1990. *Economic Analysis of Product Innovation: The case of CT scanners*. Cambridge: Harvard University Press.

U.S. Office of Technology Assessment. 1984. *Technology, Innovation and Regional Economic Development*. Washington, D.C.: U.S. Government Printing Office.

Vaughan, Roger J., Robert Pollard, and Barbara Dyer. 1984. *The Wealth of States: Policies for a dynamic economy*. Washington, D.C.: Council of State Planning Agencies.

Vernon, Raymond. 1977. *Storm Over the Multinationals*. Cambridge: Harvard University Press.

Vernon, Raymond. 1966. International Investment and International Trade in the Product Cycle. *Quarterly Journal of Economics* 80:190-207.

Von Hippel, Eric. 1988. *The Sources of Innovation*. New York: Oxford University Press.

Von Hippel, Eric. 1987. Cooperation between Rivals: Informal know-how trading. *Research Policy* 16:291-302.

Walker, Richard. 1985. Technical Determination and Determinism: Industrial growth and location. In *High-technology, Space and Society* by M. Castells, Ed. Berkeley: Sage Publication.

Weber, A. 1929. *Theory of the Location of Industries*. Chicago: University of Chicago Press.

Whittington, D. 1985. *High hopes for high-tech: microelectronics policy in North Carolina*. Chapel Hill: University of North Carolina Press.

Wilson, John O. 1985. *The Power Economy: Building an economy that works*. Boston: Little, Brown.

Wright, Gavin. 1990. The Origins of American Industrial Success, 1879-1940. *American Economic Review* 80:651-668.

Index

Economics of Science, Technology and Innovation

---

1. A. Phillips, A. P. Phillips and T. R. Phillips: *Biz Jets*. Technology and Market Structure in the Corporate Jet Aircraft Industry. 1994 ISBN 0-7923-2660-1
2. M. P. Feldman: *The Geography of Innovation*. 1994 ISBN 0-7923-2698-9

KLUWER ACADEMIC PUBLISHERS – DORDRECHT / BOSTON / LONDON

Printed in the United States
23578LVS00005B/163

9 780792 326984